ITH THE MERRIMACK RIVER
CO., BOSTON. DRAWN BY J.

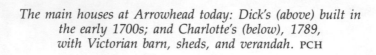

The main houses at Arrowhead today: Dick's (above) built in the early 1700s; and Charlotte's (below), 1789, with Victorian barn, sheds, and verandah. PCH

Arrowhead Farm

THREE HUNDRED YEARS OF
NEW ENGLAND
HUSBANDRY AND COOKING

BY

PAULINE CHASE HARRELL,

CHARLOTTE MOULTON CHASE,

AND

RICHARD CHASE

With Drawings By

LYNDA DIANE GUTOWSKI

THE COUNTRYMAN PRESS
WOODSTOCK, VERMONT

PHOTOGRAPHIC CREDITS

C M C   CHARLOTTE MOULTON CHASE
G C   GRACE COLLINS
C D C   CYNTHIA DAVIDSON CRYAN
P C H   PAULINE CHASE HARRELL
L M P   LIZZIE MOULTON PLUMMER

Prints from the glass plates and copies of other old photographs by
Smith/Weiler Photography, Winston-Salem, N.C.

Cover—Anonymous mid-nineteenth century painting, from a private collection.
This painting seems to have been done from the Bufford lithograph used on
the endpapers, which is a more accurate depiction of the actual landscape.

ARROWHEAD FARM was designed by Frank Lieberman.
Composition by Sant Bani Press, Tilton, New Hampshire.
Printed by Whitman Press, Lebanon, New Hampshire.
The text type is 11 point Palatino.

Library of Congress Cataloging in Publication Data

Harrell, Pauline Chase.

ARROWHEAD FARM.
Arrowhead Farm

    Includes index.
     1. Arrowhead Farm (Newburyport, Mass.)—History.
2. Agriculture—Massachusetts—Newburyport—History.
3. Cookery, American—Massachusetts. 4. Chase family.
5. Newburyport (Mass.)—Biography. I. Chase, Charlotte Moulton.
II. Chase, Richard. III. Title.
S451.M4H47   1983   974.4'5   83-10111
ISBN 0-914378-98-8
ISBN 0-914378-99-6 (pbk.)

To Glen, who loved the woods and the fields,
      and worked to preserve them;
To Charlie, who showed us the path between
      past and present;
And to all those who have cared for the land
      before us,
This book is lovingly dedicated.

Contents

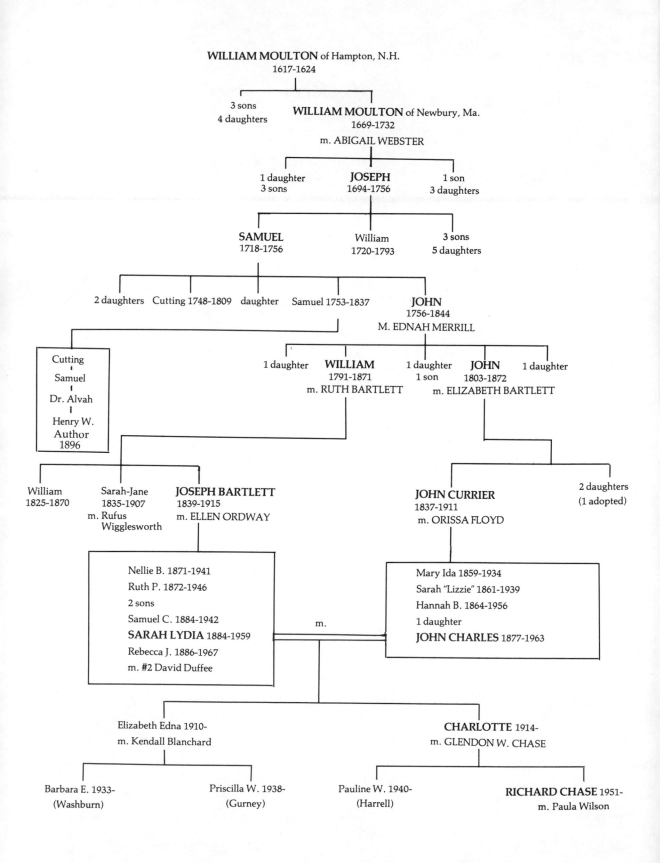

WILLIAM MOULTON of Hampton, N.H.
1617-1624

3 sons
4 daughters

WILLIAM MOULTON of Newbury, Ma.
1669-1732
m. ABIGAIL WEBSTER

1 daughter
3 sons

JOSEPH
1694-1756

1 son
3 daughters

SAMUEL
1718-1756

William
1720-1793

3 sons
5 daughters

2 daughters   Cutting 1748-1809   daughter   Samuel 1753-1837   **JOHN**
1756-1844
M. EDNAH MERRILL

1 daughter   **WILLIAM**
1791-1871
m. RUTH BARTLETT

1 daughter
1 son

JOHN
1803-1872
m. ELIZABETH BARTLETT

1 daughter

Cutting
↓
Samuel
↓
Dr. Alvah
↓
Henry W.
Author
1896

William
1825-1870

Sarah-Jane
1835-1907
m. Rufus
Wigglesworth

JOSEPH BARTLETT
1839-1915
m. ELLEN ORDWAY

JOHN CURRIER
1837-1911
m. ORISSA FLOYD

2 daughters
(1 adopted)

Nellie B. 1871-1941
Ruth P. 1872-1946
2 sons
Samuel C. 1884-1942
SARAH LYDIA 1884-1959
Rebecca J. 1886-1967
m. #2 David Duffee

m.

Mary Ida 1859-1934
Sarah "Lizzie" 1861-1939
Hannah B. 1864-1956
1 daughter
JOHN CHARLES 1877-1963

Elizabeth Edna 1910-
m. Kendall Blanchard

CHARLOTTE 1914-
m. GLENDON W. CHASE

Barbara E. 1933-
(Washburn)

Priscilla W. 1938-
(Gurney)

Pauline W. 1940-
(Harrell)

RICHARD CHASE 1951-
m. Paula Wilson

Preface

JUST AS people's birthdays have a habit of sneaking up on them, so
Arrowhead Farm's birthday sneaked up on us. We had all known, at
one time or another, that the land came into our family in the 1680s,
but somehow we hadn't put that together with the fact that the 1980s were
upon us and that meant a 300th birthday, until the fall of 1981. Once we
thought of it, a 300th birthday seemed to call for some kind of commemo-
ration. The discussions free-wheeled around traditional food, the farming
and cooking artifacts we still kept on the place, Aunt Lizzie's old photographs of farm life at the turn of the century, the way earlier generations
had farmed and lived, how Arrowhead had survived so long. And thus,
in a few short weeks, this book was born.

As we contemplated the farm's history, it seemed to us a perfect case
study of what has happened on New England farms over the past three
centuries. Different generations, it seemed, had farmed the same land differently, responding to changes in economic conditions—the growth and
decline of the overseas trade in the nearby port, new technologies in farm
machinery and transportation, changes in people's food needs and tastes.
And in each generation, we found, people had combined other occupations
and skills—shoemaking, silversmithing, furniture-making, taking in
boarders—with farming, to either bring in or substitute for cash. In these
repeated patterns of creative response to changing conditions and limited
resources lay the secret of Arrowhead's survival—and that of farming in
New England in general, with its inhospitable topography and hostile
climate. In these patterns also lay the essence of the Yankee ethic—"Use
it up, wear it out; make it do, or do without." A hard-sounding ethic,

perhaps, in these days of abundance and self-indulgence, but one that produced, from all the evidence we have at Arrowhead, a comfortable and satisfying way of life enhanced by at least a sampler of material luxuries.

In no part of New England farm life was the creative approach to the use of the relatively limited resources to produce well-being as evident as in the preparation of food. As we compared the many traditional recipes handed down to us at Arrowhead with the foods available in different generations on the farm and from the port, we were impressed by the same spirit of making the most of what was at hand. At one end of the spectrum, this meant combining vegetables, as in corn chowder, to provide complete protein without expensive or scarce meat. At the other end, it meant lavishing time and care on simple ingredients to produce delicacies like "Charlotte Rouche" or Ribbon Cake. Traditional Yankee food was rooted in traditional Yankee values.

Like most things throughout Arrowhead's history, this book has been very much a family affair. Despite the pressure of deadlines, it has been a pleasant and rewarding project for us, giving form and function to the long-practiced family pastime of exploring the contents of old photograph albums, histories, and genealogies. The three of us bring very different perspectives to bear on unravelling the mysteries of how people did things and why. Charlotte is the family archivist and curator, as well as our expert on horticulture and the natural landscape, and the most unadulterated New England cook among us. Dick is the farmer, with both a pragmatic approach to farming today, and an interest in how farming used to be done, as well as a farmer's related interest in economics, climate, and the physical sciences. Polly is a social and architectural historian specializing in the way people have created and used their environments and material possessions. The collaboration has been interesting: "That ties in with the receipts for . . ."; "Well, there was a growing demand for table crops then because . . ."; "Aunt Rebecca used to say she remembered . . ." Gradually, the pieces of the puzzle fell into place, and the farm's story took shape.

We could not have produced the book without other members of the family, past and present. Uncle William's diary from the 1850s, Aunt Lizzie's glass plate photographs recording everyday farm life in the 1890s, Aunt Ida's bulging cookbook, Charlie's reminiscences and observations passed on to two generations of children have all been indispensable.

Present members of the family have lent their efforts to the project, too. Charlotte's sister Elizabeth (Moulton) Blanchard and her husband, Kendall, have searched their memories to fill gaps in our information, as well as providing numerous recipes. Their daughter, Barbara (Blanchard) Washburn, who has long pursued the family genealogy, provided invaluable documentation; without her past sleuthing efforts, many dates and relationships would still be murky. Dick's wife, Paula (Wilson) Chase, has shared many of this generation's favorite recipes with us. Polly's step-daughter, Cynthia Cryan, has followed in Aunt Lizzie's footsteps, taking the photographs of the farm today. And Dick and Paula's son Justin, the youngest member of the family, has put up with endless conversations about the book distracting people who should have been playing with him—although he has not done so without pointing out his sacrifice! Their participation has made the project fun as well as adding to its quality, and we are grateful to all of them.

Friends have also been generous with their talents and time, and we are grateful to them as well: To Lynda Gutowski, a college friend of Polly's who has been an honorary family member since the first Christmas she came to Arrowhead and got snowbound, for her charming pen-and-ink sketches. To Jackson Smith, an excellent photographer in his own right who has a special way with old photographs, for lavishing time and skill on Aunt Lizzie's glass plates and the other old photographs from our albums to get the best possible images. To Ann Farnam, Kevin Tremble, and Dorothea Hass for reading the manuscript and making many helpful comments from their special perspectives. And to Christine Ladd and Gregg Opelka for magically transforming our cramped chicken-tracks into clean, legible prose. Without the help of all these good friends, the project would have taken much longer—and not been nearly as much fun!

P.C.H. Arrowhead, August, 1982

The family at Christmas dinner, 1976:
from left, Polly, Glen, Dick, Charlotte.

Slant-front desk, made by Charlottes's great-great-grandfather, Joseph Bartlett

Arrowhead Farm Through the Centuries

"Keep the land and the land will keep you."
OLD FARM ADAGE

THE LAND

THE FARM stand between the First Church and the Green on the High Road in Newbury is what most people know today as Arrowhead Farm. Opened in 1964, the stand is, however, a recent addition to the farm. Arrowhead Farm itself is several miles away at the end of Ferry Road, at the north end of Newburyport, and only those customers who come to pick their own strawberries usually see it. Like many farms today and in the past, its operations are somewhat scattered, but the "home place," with its three eighteenth-century houses, two barns, and other outbuildings, is still the heart of Arrowhead, as it has been since 1683.

The slant-front desk in the front room of one of the houses bulges with a bewildering assortment of deeds describing, in spidery brown longhand on fragile yellow sheets, transfers of land bounded by landmarks long since altered. For that reason, and because so many of the Bartletts, Curriers, Chases and Merrills mentioned in those deeds along with the Moultons were in fact part of our family also, we will not try here to trace who owned what when, but will speak generally of the 500 or so acres which have at one time or another been part of the farm.

Located in a sharp curve of the Merrimack River, just below the point where the Artichoke flows into the larger river, the land is quite varied in nature. Much of it is high ground with thin, gravelly soil, well-drained and frost-free early and late, but prone to drought. Other parts are low with deep, rich topsoil washed down from the high ground. This varied topography is the result of extensive geological activity during the last ice age, when great floes scoured the area, leaving kames, eskers and kettle holes in their wake. Natural stands of oak, maple, ash, white pine, cedar, and mountain laurel covered the uplands; nearer the river were marshes and meadows. Originally the land had an abundant supply of water, with natural ponds, streams, and cranberry bogs, but in the nineteenth and twentieth centuries the City of Newburyport has taken the water by eminent domain and today the farm relies on manmade ponds and irrigation from city pipes.

Most of the land is in the lee of a large hill, known as Moulton Hill. This and the particular accessability of the Merrimack along what became known as Bartlett's Cove had made this area a permanent winter camp-

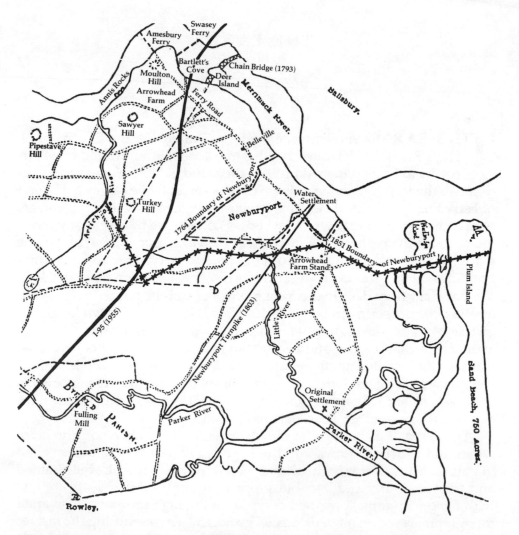

Map of the Town of Newbury showing location of Arrowhead Farm in relationship to points of local interest, based on a 1795 original, with additions

ground for the Pawtucket Indians before the plagues of the early seventeenth century decimated their population. Although most of them were gone long before the first Moulton bought the land, the blackened ring of their campfires was still visible when Charlie (John Charles Moulton, 1877–1963) plowed the river fields in the 1890s, and the number of arrowheads which his plow turned up each spring led him to name the farm Arrowhead.

Today, Ferry Road dead ends just beyond Arrowhead. For the first hundred years of its history, the farm was located on a bustling highway, as the Ferry Road led to the Amesbury ferry, and thence to points north. Indeed, for a brief time in the 1780s there was enough traffic to justify two ferries. But the coming of a bridge downriver turned Ferry Road into a little-used byway, a fact which has been significant in the farm's history.

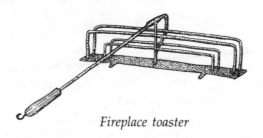

Fireplace toaster

IN THE BEGINNING

I N 1683, William Moulton (1664–1732) set out from his family home in Hampton, New Hampshire, to make a new life for himself. He was nineteen, youngest of the eight children of William and Margaret Moulton, and he had an inheritance: his father, who had died before he was born, had provided £5 in his will for the unborn child. With the proceeds of his legacy, William came to Newbury. Why he chose Newbury we can only guess. His father and mother, who had arrived at Ipswich from Ormsby, Norfolk County, England in 1635, also in their teens, had lived in Newbury for two years before moving to Hampton. By the 1680s, Newbury was a promising community, with a varied economy building on its fortunate location at the mouth of the Merrimack River. Here William settled, using his £5 inheritance to buy four acres of land not far from the ferry on which he had crossed the Merrimack.

Since most of the best land was already held by earlier settlers, the land he bought was not especially good, but its location on the main road to the northern settlements was propitious. This land became the nucleus of Arrowhead farm. Over the next three centuries, William and his descendants would buy, sell, marry into, and swap tracts of land with the Bartletts, Chases, Curriers, Merrills and other families who lived nearby.

On this land, William prospered. By the time he died in 1732, his estate was valued at £1,433, and the original four acres of land had grown to more than sixty, including salt marsh in Amesbury and Salisbury, and some land his will says was "from the last division of the Commons" in Newbury. The prototype of the Yankee Jack-of-all-trades, he is listed in various deeds as having been a weaver, trader, inn holder, and merchant. He kept a store near the ferry and built a tidewater fulling mill on the Merrimack, probably near the point known as Annis Rocks.

He may also have been a silversmith. Family tradition and the *Moulton Annals* say that he kept a black- and whitesmith shop near his store on the banks of the Merrimack and that he hired an immigrant who could hammer silver and convert coin into silver shoebuckles. There was (and is) a small deposit of silver near Annis Rocks which local legend says the Indians showed to William, but it is unclear whether he or anyone else ever mined amounts of silver worth mentioning there. His son Joseph (Joseph Moulton, 1694–1756) and numerous Moultons in succeeding generations were known as silversmiths.

William married Abigail Webster (1662–1723) and together they had nine children. He built a house on the original four acres. Captain Henry Moulton, in his charming, well-researched *Moulton Annals* of 1906, says that it was "of one timber" and that it survived well into the nineteenth century, and we think we have found traces of the cellar hole, but no picture or further description of it remains.

What was William's farm like? What crops and livestock did he raise? (Indeed, with all those other activities, what did he have time to raise?) How did the family live, and what did they eat? We must deduce the answers to these questions from his inventory and will, together with what we know of the family, the land, the town, and farming in Essex County at that time.

With many mouths to feed, but also with many hands to do the work, he undoubtedly did farm extensively, despite all his other interests. In the 1680s and 90s, probably only the economy of Boston could support many merchants or artisans who did not also farm, and Newbury was far from Boston. The town, which at that time included all of what is now Newbury, Newburyport, and West Newbury, had about 700 inhabitants in 1688.

Because of the abundance of meadow and marshland, livestock had been important in Newbury from the time of its settlement in 1635, and by 1695 there were more than 5,000 sheep in the town. They supplied wool for two fulling mills in the Byfield section, as well as the one William Moulton started at Annis Rocks. There were also many cattle, and just downriver from William's land, a tannery at Bartlett's Cove processed hides. Slightly upriver, on the Artichoke, Sgt. Emery ran a corn mill, which the town had given him twelve acres of land to build in 1678. Wheat had never done well in the area, but corn, rye and barley were important crops, both for local use and for export.

The river and the ocean played a large part in the local economy. From the very beginning, fishing had provided food and trade commodities. By the time William arrived, ship building and trading were becoming important. A number of docks had already been built near the center of the "waterside" area, and in 1684 the General Court appointed a customs officer for Newbury. From here, merchants carried on a brisk trade with the West Indies, shipping out clapboards, barrel and pipe staves from local and upriver woods, grain, salt pork and beef from local farms, and salted and dried fish from the river and ocean; and bringing in sugar, molasses, and other goods. Further upriver were numerous shipyards, which by 1714 had launched more than a hundred vessels. By 1710, Tom Barrett was running a shipyard at Bartlett's Cove. It is hard to imagine that William and his neighbors were not selling wood to him and other shipwrights, as well as supplying the waterside merchants with goods for export.

Despite all this trading, the family farm was, by our standards, almost completely self-sufficient, producing not only most of the food the family ate, but much of their clothing, tools, and furniture as well. The men and boys cared for the livestock and field crops, and made furniture, tools, and leather goods, while the women and girls tended the poultry and the

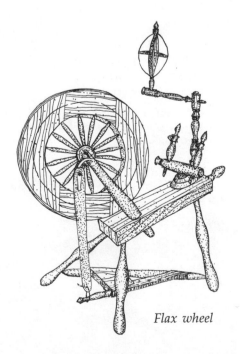

Flax wheel

kitchen garden, made cheese, butter, candles, and soap, and turned flax and wool into cloth and clothing. The typical farm of this era produced corn, rye, oats, and barley for man and beast; peas, squash, beans, pumpkins, turnips, and other root vegetables; and flax for cloth. The apple orchard provided apples for cider and the winter staples of pie and sauce. A few cows supplied milk, cheese, and butter, and oxen pulled the plow; eventually these would furnish tough, stringy beef, tallow, and leather. Sheep for wool and meat; hogs for salt pork to eat and trade; barnyard fowl for eggs and meat; and horses to transport people and freight rounded out the livestock.

Tools were simple. Spades, hoes, forks, and rakes were made entirely of wood; even the plowshare was wood. Sledges and heavy two-wheeled carts, each wheel a slice of tree trunk with the bark removed, were the only vehicles.

The family also harvested and made use of many things around them which nature had provided. From their own woodlot they cut firewood, wood to sell to shipyards and the West Indies trade, and the raw materials for the farm buildings, furniture, and tools. From the maples they collected sweet sap to make into maple syrup. From the marshes, they cut salt hay to feed the livestock through the winter. From the river came fish: shad, sturgeon, tom cod, and salmon. (The latter were so plentiful in the seventeenth century that local apprentices had it stipulated in their contracts that they would not be forced to eat salmon more than six times a week.) And at different seasons, the land provided tasty additions to the table: wild strawberries from a sunny meadow near the river; watercress and peppermint from the hillside spring; blueberries—low bush from the hillsides and high bush from the swamp areas; cranberries from several bogs.

To these were soon added sweet, pungent molasses and sugar from the West Indies, and tea, spices, and raisins from the expanding overseas trade. In these available foods, we can see the beginnings of New England cooking as we have known it over the centuries: corn bread, "Rye and Injun," brown bread, Indian pudding, baked beans, apple, squash, and cranberry pies.

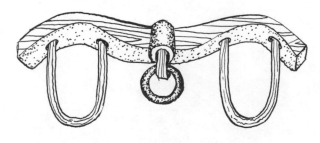

Ox yoke

A cooking fireplace of William's era

Although we have no evidence of what William's house looked like, we can guess much about the kitchen—or hall, as it was called—from other surviving houses of the period, and we can picture the family in it. It was the principal room of the house, where they spent most of their time.

The huge brick fireplace dominated the room. It had ovens in the back, and the women had to be careful that their skirts didn't catch fire when they tended the baking. Iron cranes held large iron pots which could be hung near or far from the fire on adjustable hooks, called trammels. There was space for more than one fire, too, of different sizes and different kinds of wood for simmering, roasting, or searing. In cold weather, much of the heat from the fire went directly up the broad-throated chimney, for fireplaces were a relatively recent technological development in the seventeenth century, and their design was still primitive. The spinning wheel stood close by; there was always spinning or sewing for the women to do in any spare moment. Near the fireplace, too, benches with high backs provided a cozy, draft-free spot to sit. Here the men might whittle small tools and utensils of an evening, or read to the family from the Bible.

The trestle table—made, like most of the furniture, from wood cut nearby—served as a cooking counter and work table as well as dining table. Presses or cupboards held wooden, earthenware, and pewter eating utensils. In the winter strings of onions and piles of squashes and pumpkins lined this, the only room both dry and safe from frost. Just off the kitchen was the buttery, a cool room where milk set in pans waiting for the cream to rise. Here the women made and stored butter and cheese. Somewhere nearby was the entrance to the root cellar, where apples, as well as turnips, parsnips, and carrots were stored, cold and damp but safe from frost. Store-rooms in the leftover space under the eaves in the attic held chests and barrels of meal and flour; and from the rafters hungs strings of dried apples, pumpkins, and perhaps other fruits, and bunches of dried herbs.

William had at least one servant, an Indian woman. Under his spon-sorship, she joined his church. The handwritten records show some crossed-out attempts at spelling her Indian name, but she is finally listed as "Dinah." While few Indians survived in the immediate area after the plagues in the first years of the seventeenth century, a few evidently lived near the back country of the Artichoke for many years. Newburyport novelist John Marquand noted in his essay on the history of Curzon's Mill that his great-aunt remembered seeing an occasional few silently paddling birch bark canoes down the Artichoke into the Merrimack as late as 1830. This tiny remnant was generally quite meek, but in 1695, in one of the recurring French and Indian wars, a raiding party carried off nine members of John Brown's family from their house near Turkey Hill, less than three miles from William's house. Most of them were recovered the next day, but at least one boy was killed. And one of Abigail's cousins was Hannah Dustin, who lived a few miles upriver at Haverhill. Captured by Indians who killed her week-old baby during a bloody raid on that settlement in 1695, she escaped six weeks later, killing and scalping ten Indians in the process and lived to tell her horrifying tale. We can imagine that the cozy farm kitchen on the Ferry Road sometimes seemed isolated and vulnerable in the 1690s.

By 1729, as the map shows, a sizeable farm community had developed in the area, providing plenty of social and political activity. Each new house-raising or barn-raising created an occasion for the neighbors to get together to help the owner, and then celebrate around a bountiful board. Neighbors helped each other with plowing, haying, cornhusking and other tasks as well. We don't know where the school was then, but William and his neigh-bors shared the Puritan belief in education, so there must have been one, even if only a "dame school" in someone's kitchen.

Although we have a description of a Bartlett ancestor mounting his horse and riding furiously to Boston to participate in the overthrow of the hated Governor Andros in 1689, Colonial politics do not seem to have in-volved William's family much. Local politics, however, were another mat-ter. Complaining of the difficulty of travel to the original meeting house, William and several neighbors formed their own parish in the 1690s near Sawyer's Hill. After several years of conflict with the town meeting over their right to do this, a compromise was reached, and a new parish was

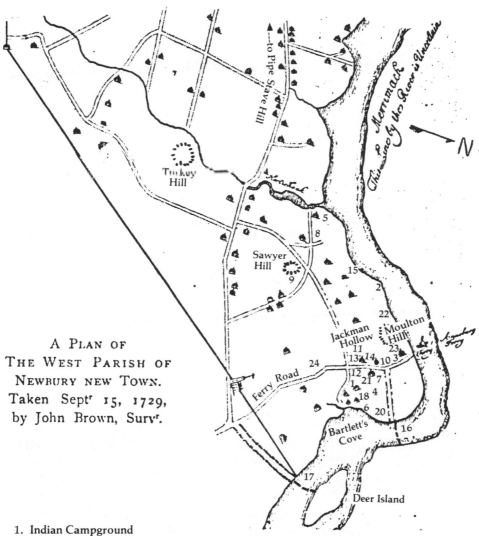

A PLAN OF
THE WEST PARISH OF
NEWBURY NEW TOWN.
Taken Septr 15, 1729,
by John Brown, Surv^r.

1. Indian Campground
2. Annis Rocks
3. William's House Site
4. The Tannery
5. Sgt. Emery's Mill (later Curzon's Mill)
6. Tom Barret's Shipyard
7. The Boiling Spring
8c. Cranberry Bogs
9. William's Church
10. Early 18th Century House
11. Original Site of 1789 House
12. 1873 Site of 1789 House
13. Present Site of 1789 House
14. 1790 House

15. Site of Samuel Moulton's House
16. Swasey Ferry
17. Chain Bridge (Essex Merrimack Bridge)
18. Site of Great Grandmother Bartlett's House
19. Site of Great Grandmother Bartlett's School
20. The Point Field
21. Spring with Hydraulic Ram
22. The Laurels
23. Site of Moulton Castle
24. Will's Woodlot / Sandy Acres

Map of the area immediately surrounding Arrowhead,
based on a 1729 original, with modifications

established farther west, near Pipe Stave Hill. The "renegades" in William's immediate neighborhood seem to have continued their own little parish, however, which may have had something to do with the fact that this area eventually stayed in Newbury when West Newbury broke off as a separate town. Its economic ties were more and more "down river."

Of that thriving community formed by William Moulton and his neighbors, not one of the houses shown on the 1729 map remains today. What happened to them? A few survived until the end of the last century; most had disappeared a hundred years ago. Of some there is not a trace today. Of others, the practiced eye can read the story in the landscape: a softly rounded cellar hole, a clump of lilacs, the faint parallel indentations of a cart path. Like the blackened campfire rings and the arrowheads of the Indians, the traces of these long-departed inhabitants still mark the land.

BUILDING TO LAST

AS THE land is our link to the first generation at Arrowhead, the three houses that stand on the farm today are our link to the eighteenth century. Growing up in an old house, you become intimately familiar with its textures and idiosyncracies: with the wide pine floorboards that slope toward a settled corner, with the worn treads of the steep stairs, with the bubbles and ripples in the small windowpanes. It's hard to escape an occasional feeling of closeness to the people who lived there, too: those people who wore the familiar hollows in the stair treads, spilled the ink that still stains the desktop, and sat in the same Windsor chairs around the fireplace two hundred years ago. Yet, for all our sense of familiarity with these houses and their occupants, many of the details of who built them and when are hazy.

Sometime before the middle of the eighteenth century, one of William's sons or grandsons built the big foursquare Georgian house on the brow of the hill that led down to the ferry. Family tradition says that he framed it up near Annis Rocks, then moved the frame and finished the house on this site to be near his father after the death of a brother. Some evidence suggests this might have happened in 1717–18; other records seem to indicate that it was in the 1740's. The same family tradition holds that the ell which now serves as the Arrowhead Farm office was the original Moulton family silver shop, moved across the road and added to the house after William's son Joseph moved his silver and goldsmithing business downtown, and structural evidence seems to bear this out.

The oldest house at Arrowhead dates from the first half of the eighteenth century

Originally built with a massive center chimney, this house has been much modified over the years. The major change seems to have occurred at the end of the eighteenth century, when it was apparently divided for two branches of the family. The center chimney was removed and replaced with a gracious hall and stairway, and chimneys were added on either side, each serving a parlor to the front and a cooking fireplace with a wall oven in the kitchen to the rear. In doing this, the remodellers neglected to replace the structural function of the massive center chimney, with the result that, when Dick and Paula took over the house in the 1970s, experts called in to look at the foundations concluded that it was being held up by "nothing except habit." Although that habit had sufficed for well over a hundred years, we decided that the time had come to augment it with some steel jackposts. Since then, the center of the house has gradually been raised six inches.

The second house has also been much altered and moved. A center-entrance, transitional Colonial/Federal-style house with "ship stairs" winding up from the front hall, wide pine wainscotting and simple Federal mouldings and fireplaces, it was built in 1789 in Jackman Hollow (see map). In 1873, when John Currier Moulton (1837–1911) and his wife Orissa (Orissa Floyd Moulton, 1835–1915) were raising their family in this house, she, we've been told, felt lonely in that isolated spot. So, with twenty-four yoke of oxen, they moved the house to a hill on the Ferry Road near the other two houses. Probably at this time, the rear lean-to portion was removed and replaced with a more modern kitchen area.

In 1897, when the city took over the land on the river side of Ferry Road for a city water supply, the house was again moved, to its present site. This time, one horse and a winch attached to a series of trees did the job. Charlie and his sisters often reminisced about living in the house during the two days while it was being moved, and Lizzie (Sarah Elizabeth Moulton Plummer, 1861–1939) was particularly fond of referring to it as the night she "slept in the gutter." With this move, the house was renovated once more: several feet were added to the rear and milk rooms, sheds and privies were rebuilt to connect the house with the barn. A side porch on the south and a wide verandah and covered piazza wrapping around the front and breezy north side of the house also date from this period. Today, the house is a pleasant melange of the eighteenth and nineteenth centuries, with a few nods to the twentieth in the form of heat, plumbing, and modern appliances.

The third house sits where it was built in 1790 and has had minimal alterations. It was, we know from family tradition, a direct copy of the house built the previous year in Jackman Hollow, and comparison of the two corroborates this. The ceilings are much lower, however, bearing out the story that, while it was being built, a severe gale blew down the frame. Rather than buying or cutting new timbers, the housewrights cut off the damaged portions and reused them—"Use it up, wear it out; make it do, or do without." The low ceilings, combined with the very primitive country versions of Federal detailing, give this house a special charm.

Of life at Arrowhead in the eighteenth century, we know fewer details than of William's generation, oddly enough. From mid-century inventories, we learn that in addition to flax, wool, and other items William had produced, his son Joseph (1694–1756) and grandson Samuel (1718–1756) introduced to the farm two important new crops, potatoes and English hay. Potatoes were first brought to New England in 1719 by some Scotch-Irish families who settled in southern New Hampshire and western Massachusetts. Although potatoes became a staple by the end of the century, they caught on fairly slowly at first, and their presence in the 1750s indicates that Newbury in general and Arrowhead in particular were thoroughly up-to-date agriculturally.

English hay also made its appearance in the first half of the eighteenth century. Timothy, or herdsgrass, as it was also called, seems to have developed from a variety originally discovered near the Piscataqua River. Popular in England, it was reintroduced to New England, where it became a vital crop. It prospered on cleared land, and livestock thrived on it. Its introduction freed farmers from dependence on the natural meadows and salt marsh to support cattle, sheep, and horses, and the numbers of these increased greatly. Interestingly, large quantities of beef are also mentioned in both Joseph's and Samuel's wills. It seems likely that they were selling

*The 1789 farmhouse as it looked on its second site, in the 1890's (top);
the same house after the move to its present location on the opposite
side of the road in 1897; (bottom) the 1790 house, foreground, with
the earlier house in the background.*

some to the port for the growing West Indies trade, along with potatoes, lumber, grain, and other farm products.

This growing trade, and generally increased standard of living, brought a greater variety of foods to the table. Tea from England's East India Company and coffee from Surinam were readily available, along with molasses and sugar and more kinds of spices. Raisins and an occasional lemon or orange came to the port from Spain and Portugal and perhaps made their way upriver this far. The variety of vegetables and fruits grown at home increased, too—cabbages are mentioned in Joseph's inventory, and we know that salad greens, cauliflower, asparagus, and melons of various kinds were being grown in the area.

Sarah Anna Emery, in her *Reminiscences of a Nonagenarian*, about growing up in Newbury at the end of the eighteenth century, wrote of the garden around her mother's house, not far from Arrowhead:

> *Before the house stretched a large garden, well-stocked with pear, peach, and cherry trees. Currant and gooseberry bushes grew luxuriantly beneath the sheltering board fence that separated the enclosure from the broad fields and orchards around. There was a clump of quince bushes in one corner, and in another two Plum Island plum bushes that had grown from stones taken from fruit brought from the island. There was also a great variety of medicinal and sweet herbs, and from early spring till late in autumn the borders on either side of the gravel walk were gay with flowers.*

Given the proclivities of later generations of the family, it is easy to conjecture that cultivated flowers had made their appearance at Arrowhead by this time, too.

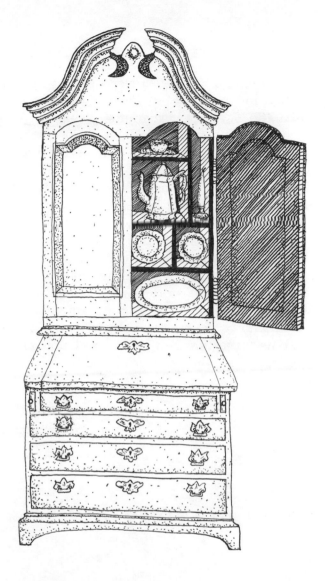

Fireplace wall in the 1790 house; and, to the right, the late 18th century bonnet-type secretary

Material comforts generally increased in this century. The fireplaces grew smaller and improved in design so that less heat went directly up the chimney. The ovens moved to the front, a boon to women's skirts. With tea came imported cups and saucers—delicately painted Bristol ware and blue Staffordshire—and coin silver spoons, made by the Moulton silversmiths, to enhance its enjoyment. These and several pieces of hand-some locally made furniture from this era, together with the houses themselves, suggest a comfortable style of life and a pleasant level of amenities.

Otherwise, life changed slowly. The French and Indian Wars continued until 1763, but local Indian raids were a thing of the past. A few friends and relatives joined the expedition which captured Louisburg in 1745, but for most the wars remained remote. Local politics revolved around the creation of new parishes and the provision of schools for children. The mercantile settlement by the waterside grew and prospered with shipbuilding and trade, and the 1760s reverberated with hot debate about its petition to become a separate town. In 1764, over the protests of the rest of the town, the General Court granted a charter, and Newburyport was created. Despite the apprehension, the major discernable effect on the farming community upriver seems to have been a continued expansion of markets for "country goods."

The Revolution came fast on the heels of Newburyport's incorporation. We know very little about how the war affected life at Arrowhead. Family tradition tells how the news of the Battle of Bunker Hill travelled, as news often did, across the marshes, shouted from one haying crew to another, "a whoop and a holler" distant. A Revolutionary musket has also been handed down, along with the image of an ancester shouldering it and marching off to Boston. (And no detailed history of the Revolution is as effective as fingering and shouldering that musket, as every Arrowhead child knows.) Samuel's son John (John Moulton, 1755–1844) did enlist briefly, but for the most part, as in wars before and since, the family seems to have stayed at home and produced food instead of going off to fight.

GREAT-GRANDMOTHER BARTLETT'S ERA

THE YEARS after the war were hard ones for Massachusetts farmers. The Continental currency collapsed, taxes were high to pay off war debts, and the hard-pressed merchants in turn pressured farmers to pay debts incurred during the war in cash they didn't have. A series of harsh winters made matters worse, and, while Essex County was not in as desperate straits as the western counties where Shay's Rebellion ensued, times were lean here, too.

With times hard, and the family growing, it is little wonder that, like many others in the older farm communities in this period, several members of the family left Newbury for greener pastures. In 1786 Samuel's sons Cutting (Cutting Moulton, 1748–1809) and Samuel (Samuel Moulton, 1753–1837) removed to the new town of Parsonsfield, in the District of

Maine. We have a description of how, on December 15 of that year, they bundled their sixty-five-year-old mother into a sleigh on which they had constructed a tent of bedposts and quilts to keep her warm and set out for their new home. While mid-winter may seem to us a strange time to begin such a venture, it was in fact the easiest time to travel over land, when the frozen rivers became smooth highways for sleighs and sleds. Moreover, it was the "slack time" for farmers, and much travel seems to have taken place then.

Two years later their uncle William (William Moulton, 1720-1793), a silversmith, and his wife Lydia moved with their family to Marietta, then being founded in Ohio in the newly opened Northwest Territory. The annals of the town contain a glimpse of that family we have always cherished: during an Indian raid in 1791 when the townspeople gathered in the "safe house," William, aged 70, came

> . . . with his leather apron full of old goldsmith's tools and tobacco. Close at his heels came his daughter, Anna, with the china teapot, cups and saucers. Lydia brought the Great Bible. But when all were in, the mother was missing. Where was mother? She must be killed. "No," says Lydia; "Mother said she would not leave the house looking so. She would put things away and a little more to rights, and then she would come." Directly Mother came, bringing the looking glass, knives and forks, etc.

Thus did New England habits move West.

Those who remained on the family homestead were about to witness the most exciting event in Ferry Road history: the passage of George Washington on his tour of New England in 1789. A housewright building the house in Jackman Hollow was moved to record the great event in chalk on the attic rafter where he was working and where Charlie would find his words in the 1920s: "Mr. Washington is here." From a window of the old house on the brow of the hill leading down to the ferry, John's young bride Ednah (Ednah Merrill Moulton, 1767–1852), it is said, got a clear view of the great man as his entourage passed. And the neighborhood talked for generations of how the General stopped to pick up and gently kiss a Bartlett baby. When the procession reached the ferry,* the river was crowded with craft of all descriptions, including American and French men-of-war, which greeted Washington with a 21-gun salute. A memorable sight indeed for the stolid citizens of Ferry Road!

---

*In 1789, there were two ferries: the original one at the end of Ferry Road, and a new one begun in September of that year by Joseph Swasey, reached by a lane which branched off the main road (see map). Washington's diary says that he crossed the new one.

Roof rafter in the 1789 house; and (opposite) the Chain Bridge in the 1890's, replacing Timothy Palmer's 1793 span.

Less than a decade later, another event occurred along the river which, although far less dramatic than Washington's visit, had a far greater impact on the future of the neighborhood. In 1793, Timothy Palmer built a bridge across the Merrimack at Deer Island. From a main highway north, the Ferry Road became first a secondary road, and then, when the ferry was discontinued a few years later, a dead end. To this turn of events more than any other, perhaps, we owe the preservation into the twentieth century of old things and old ways at Arrowhead.

In the booming years of the 1790s and early 1800s, when the Napoleonic wars abroad created an increased demand for American foodstuffs, Newburyport reached its peak of prosperity. Timothy Palmer's bridge and the ambitious Newburyport Turnpike to Boston, built in 1803 with Newburyport money and still the straightest road on the map of eastern Massachusetts, attest to the town's wealth and initiative. The handwriting was on the wall, however; already under construction was the Middlesex Canal, which would divert the upcountry trade from its natural outlet at the mouth of the Merrimack to the larger port of Boston. Soon after this canal opened in 1803, Jefferson's Embargo of 1807 devastated all of New England's trade, and in 1812 came war with England, similarly disastrous. As if this weren't enough, in 1811 the greatest blow of all fell on Newburyport—the Great Fire which destroyed the town's commercial center. Although much of the downtown was quickly rebuilt and some merchants made fortunes in privateering during the war, these combined calamities would end forever Newburyport's status as one of New England's leading seaports. Henceforth, it would function as êntrepot for a small, isolated corner of northern Massachusetts. Newburyporters would trade with the West Indies until the 1870s, and build some ships into the 1890s, but the great days were over. And that inevitably affected the farms along the old Ferry Road.

We catch the barest glimpses of life at Arrowhead at the beginning of the nineteenth century from the girlhood memories of Charlotte's "Great-Grandmother Bartlett" (Elizabeth Bartlett Moulton, 1800–1897), recounted to her young grandson, Charlie, in the 1880s and 90s. A petite woman with somewhat elegant tastes, if we are to judge by her ivory silk wedding dress, dainty lace-trimmed blouses, and elaborate bonnets in the attic, Great-Grandmother Bartlett is an important link in our chain of family history. She seems to have played the same role for Charlie and his generation that he played for us: living with the family to a ripe old age and passing on stories gleaned from a lifetime of interest in how things change and how they used to be. She and her numerous brothers and sisters had been orphaned at an early age, and family tradition has always linked her parents' deaths to an accident in which their sleigh plunged through the ice in the river. The "unfortunate orphans," as the records call them, seem to have been well provided for, however. Their father is listed as a carpenter and housewright, and various records suggest that he built both the house in Jackman Hollow and the cherry slant-front desk in its front room.

Great-Grandmother Bartlett's first trip "downtown" to the port, she always said, was to view the ruins of the Great Fire, when she was twelve. While her father, like Sarah Anna Emery's nearby, may have travelled to town as often as once a week to sell eggs and cheese, it does not seem

that the life of Ferry Road families revolved around activities in the port at that time. Her other memories, such as that of trudging to a neighbor's house for live coals whenever the fire went out during the night, suggest a life unchanged in many ways from the previous century.

On the other hand, receipts for several wagons and chaises in the early 1800s suggest that the family was keeping up with fashion in transportation in these years. They also continued to acquire other comforts of life from Newburyport, as well: lovely furniture in the new Hepplewhite and Sheraton styles, delicate silver spoons made by their silversmith cousins, china from England, and imported foodstuffs such as cinnamon and nutmeg from the Far East and chocolate from South America, processed by the new Baker's factory in Dorchester. It seems probable, from this fairly affluent standard of living, that the farms in the area had come through the hard times after the war, and were now sharing in the port's fleeting prosperity by selling foodstuffs for export and to feed the artisans and merchants.

We know too that, like many farm families in New England, they took education seriously. Although we have found no mention in the records of a school in this area, Great-Grandmother Bartlett said she went to a school near William Moulton's old church, and we have several books published in the 1790s and 1800s inscribed, "Elizabeth Bartlett: her book," including a practical medical treatise and a five-act play. Presumably, the men in the family were at least as well educated as the women. Indeed, visitors to New England in these days often commented in amazement at the uniformly high educational level among farm families, and the collection of books, almanacs, and treatises handed down to us bear this out.

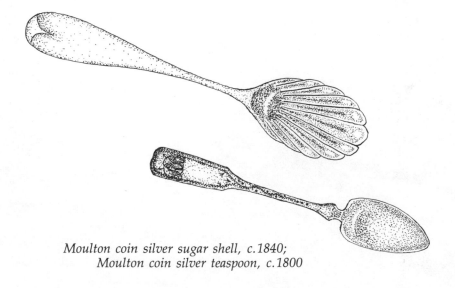

Moulton coin silver sugar shell, c.1840;
Moulton coin silver teaspoon, c.1800

CHANGING MARKETS, CHANGING WAYS

ALTHOUGH life at Arrowhead seems to have changed very slowly, in New England as a whole significant developments were coming in a rush in these years: the Essex-Merrimack Bridge at Deer Island and the Middlesex Canal were only the first of countless such transportation improvements across New England. Factories would soon spring up at the falls of the Merrimack, as they already had along the Charles and Blackstone Rivers, and towns would burgeon around them. And before the middle of the century, railroads would push through the countryside, linking farm and town. Yankee ingenuity was developing new tools to revolutionize farming, and agricultural improvement societies were introducing new crops and new breeds of livestock.

By patching together an assortment of receipts, letters, books, pictures, family possessions, and traditions from that period, and a diary and account books kept by Uncle William (William Moulton, 1825–1870), with what we know of New England farming practises and the regional and local economy, we can get a fairly good picture of nineteenth century life at Arrowhead.

We know, for example, that one of the earliest and most important links between farming and manufacturing in Massachusetts was shoemaking. A well-worn cobbler's bench and many old shoemaker's tools suggest the importance of shoemaking to early generations at Arrowhead. From the earliest days, tanneries such as the one at Bartlett's Cove had processed hides from local farms, and farmers and their families had made shoes for themselves and their neighbors during the winter. Spurred on by the drive to increase "domestic manufactures" during and after the Revolution, shoemaking became a vital cottage industry in the Massachusetts economy by the early 1800s. Shoe manufacturers furnished the stock; farm families cut out the uppers and pegged them to the soles, then returned them to be finished and packed in factories. By 1855, Massachusetts turned out 33,000,000 pairs of shoes and 12,000,000 pairs of boots each year, many still in this piecework manner.

Uncle William's 1858 diary tells of picking up stock in Amesbury, working on the shoes at odd moments, and returning the finished products, mostly in the winter months. Despite the mechanization of much of the industry, this occupation continued to provide welcome cash to Arrowhead, as to many farms, throughout the nineteenth century. We still have the pair of shoes John Currier Moulton made for his wedding in 1857, and very fine work they are, too. As late as 1900, Sarah (Sarah Lydia Moul-

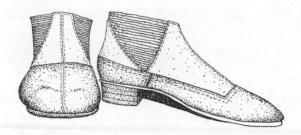

ton, 1884–1954), did beadwork for women's shoes from Haverhill, as did many girls of her generation, and well into the 1930s Charlie resoled the family's shoes whenever they needed it.

While farms around Boston became increasingly specialized in the first half of the nineteenth century, producing table crops and dairy goods for the city dwellers, Uncle William's diary reflects a continuation—perhaps even an intensification—of the "Jack-of-all-trades" pattern of farming around Arrowhead. He talks of planting corn, beans, potatoes, peas, barley, oats, and rye, and of haying the salt meadows. But he also talks of shoemaking, renting out his oxen to others for special hauling jobs, selling "sticks" (masts) to the shipyard, making an ox yoke and an axletree for his horse rake. Receipts and account books also show that in this century some of the women did sewing for people in town, and that they produced eggs, butter, and cheese for sale there, too. Other men in the family worked part-time in boatyards, or hired out their teams and themselves to haul banks dories from the Amesbury boatyards to the Gloucester fishing fleets. Interestingly, however, we find no record of anyone in the

Hauling dories was one source of income. John Currier Moulton's wedding shoes, (opposite) which he made himself in 1857; and cobbler's bench

family working in the factories that were so rapidly growing up in New-buryport, as in the rest of the Merrimack Valley.

As in earlier generations, some members of the family sought economic opportunity elsewhere. In the 1840s, Aunt Jane Moulton's (1835–1907) husband, Rufus Wigglesworth, (1826–1900) went to sea as a ship's cook, rounding the Horn with Gold Rush passengers and eventually making voyages to most of the world. His small deer-hide trunk, full of colorful shells from the South Pacific, a tiny Chinese slipper, and rock samples from distant shores, has seemed truly a treasure chest to each generation of children exploring the attic on a rainy day. Rufus and Jane continued to live in the old house on Ferry Road, and it seems likely that, in addition to the silks, lacquered fans, and other little luxuries he brought back, his income was a welcome addition to the farm.

Others, like Cutting, Samuel, and William before them, moved away permanently. In the 1870s, Orissa's brother, Herman Floyd (1849–1906), emigrated to Kansas, where he became mayor of the brash new cow town of Abilene. Charlie always vividly remembered the times this dashing frontiersman visited Arrowhead, and the metal-tipped Sioux arrows he brought took an honored place beside the stone arrowheads Charlie had found on the home place. Another brother of Orissa's moved to Washington, D.C. Through such ties, as well as through books, newspapers, and magazines, those who remained on the farm kept in touch with the rapidly changing world around them.

Uncle Rufus's treasure trunk

Uncle William's diary mentions fixing his horse rake. He also tells of a neighbor who had lost over $100 worth of "irons" when the load on his gundalow shifted while he was bringing his equipment upriver from haying on the marshes. By the 1850s, then, iron tools were clearly part of life at Arrowhead—a dramatic change from the earlier generations. Until the early nineteenth century, farming methods had changed little since the days when the first William settled in Newbury. Wooden hand tools, heavy wooden-wheeled carts, and even clumsy wooden plows were the farmer's only aids. But, beginning in the 1820s, the new factories soon turned out a bewildering variety of ingenious devices to ease the farmer's every task: iron plows, horse-drawn rakes and mowing machines, corn choppers, iron wheel rims for lighter wagons, iron spades, forks, and rakes. They were,

Charlie mowing in the Point Field. LMP

of course, expensive, requiring large farms to make them really worthwhile. And especially the early machines were fragile and thus unsuited to the rough, rocky New England land. Nevertheless, some of the new machines were making life easier at Arrowhead.

While John Currier could collect twenty-four yoke of oxen from neighboring farms to move his house in 1873, better, lighter-weight equipment meant less use of oxen and more use of horses as the century progressed. With heavier use of horses on the farm and also in the city, the demand for hay and feed grains such as oats and barley increased, and it is not surprising to find William planting these crops.

There are other indications that progress in farming techniques was reaching Arrowhead. Uncle William's diary mentions attending a fair on Broad Street in 1858, with a pulling contest; and an assortment of prize ribbons and certificates tells us that the people at Arrowhead also regularly attended and competed in the Amesbury fair. Well-thumbed copies of farmers' almanacs from early in the century, and of magazines like the *New England Farmer,* and manuals on nearly every aspect of farming from later in the century show that the farmers in the family were eager for the latest information. These publications and the agricultural fairs, sponsored by the local agricultural improvement societies, told them about the many exciting new developments in livestock: Jersey cattle, imported in the 1850s, to improve the butterfat content of milk; Holsteins, brought in a few years later, to increase the quantity of milk produced; new breeds of poultry to improve the quality and quantity of eggs. Many of these new types of poultry were originally brought in by sailors who picked them up in the Orient and kept them on board ship for fresh eggs, and we wonder if Uncle Rufus Wigglesworth might have added to Arrowhead's flock in this way.

Photographs of Arrowhead in the late nineteenth century show hen pens and barns of the type being advocated in the journals at mid-century. These were better designed, and provided warmer shelter in winter, thus contributing to the quantity of eggs and milk and lengthening the season of production. The journals also recommended better feed, and it seems reasonable to think that much of the hay and grain Uncle William mentions planting and harvesting was also of the new, improved strains and helped produce healthier livestock.

Arrowhead sold eggs, milk, butter, and cheese downtown. By mid-century, Newburyport's economy had changed greatly from the halcyon days of shipping. The shipyards were still turning out ships, many of them graceful Clippers for the China trade, but that trade had moved to Boston and New York. One or two merchants still imported molasses from the

Hen pens at John Currier Moulton's place in the early 1890's, and across the road after the move, Charlie with a Barred Plymouth Rock hen in the lower right. LMP

West Indies for the local market and the Caldwell's rum distillery. But manufacturing seemed the wave of the future. In 1851, when the town became a city and incorporated the area up to and including the bend in the river, two or three new textile factories were already in existence or planned in the downtown. The railroad had finally reached Newburyport in 1847, and hopes were high for a prosperous industrial future. Like many rosy futures, this one never materialized. The few factories which had been built remained, however, their operatives providing a market for local farms, which partially compensated for the loss of the export trade.

Sheep, which had been so important in the beginning at Arrowhead, are not mentioned anywhere in our nineteenth century sources, nor is flax. Sheep raising declined in general in New England in these years,

driven down by western competition, and with the growth of new factories producing cheap cotton cloth, flax virtually disappeared as a crop.

Uncle William does mention planting potatoes in large quantities. We
know from other references that they were a major crop at Arrowhead
in the second half of the century; in fact, the land is particularly well suited
to potatoes, and many generations have grown them. The same textile
factories which had eliminated the need for flax now provided a market
for starch. He also mentions planting beans, another staple crop, which
were shipped by train to the new factories which canned them.

In addition to the cash crops, the farm continued to produce most of
the food for the family in this era. Corn—some of the old flint corn to be
ground for meal, some probably of the new sweeter varieties for eating
on the cob, and much for fodder for the cows—remained a major item.
Root crops such as turnips, carrots, onions, and potatoes which stored
well, cabbage, and the many improved varieties of squash were still
mainstays, but the kitchen garden provided a wide variety of seasonal
vegetables as well. Apple, peach, pear, and cherry trees, quince bushes,
and a bed of the new improved cultivated strawberries provided fruit.

In the kitchen, iron ranges and lightweight cookware gradually replaced the old fireplaces and their heavy iron pots. To start the fire, sulphur
matches from the store replaced the tinder box or live coals borrowed from
a neighbor. Kerosene lamps replaced candles and whale oil. By the end
of the century, if not before, ice chests, cooled by ice cut from local ponds
and stored insulated with sawdust, helped keep foods from spoiling during
the hot months. The glass jars for preserving fruits and vegetables by canning, introduced by John Landis Mason in 1848, increased hot work in
the summer but provided a greater variety of foods throughout the year.
Ingenious new grinders, choppers, churns, and other mechanical devices
made many tasks easier. And the house on the hillside near the river had
water brought up to it from a spring down the hill by a hydraulic ram.

These changes were not made without some regrets. Many people com-
plained that food just didn't taste good cooked on a range. And Harriet
Beecher Stowe probably spoke for many when she wrote,

> *Would our Revolutionary fathers have gone barefooted and bleeding over*
> *snows to defend air-tight stoves and cooking ranges? I trow not. It was*
> *the memory of the great open kitchen-fire . . . that called to them through*
> *the snows of that dreadful winter.*

What the farm kitchen lost in cheeriness with the open fire, however, it
undoubtedly made up in the delights of newly available manufactured
goods: inexpensive, brightly printed cloth, tinware and enameled steel

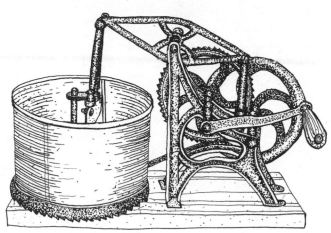

Kerosene lamp on cast iron wall bracket; and a 19th-century food chopper. When the handle is turned, blades go up and down and the container rotates.

cookware, a wide variety of crockery and china, sparkling pressed glass in charming patterns. Flyscreens for the windows added immeasurably to summer comfort in these years, and the creak-slam of the screen door became one of the distinctive sounds of summer. With the new conveniences, there was more time for little amenities, and potted geraniums and other plants made their appearance in the kitchen at this time.

The new improvements cost money, however, and cash was hard to come by. The factories in Newburyport could not replace the profitable foreign markets; the farming area around the city was less prosperous in general than it once had been. And despite the arrival of the railroad, the lucrative Boston market was too far away to affect farmers here. So the old ways hung on at the end of the Ferry Road.

John Marquand once ruminated about his elderly great-aunts' style of life at Curzon's Mill at the turn of the century, which he believed was far more reflective of the late Federal era they had grown up in than of the time in which they were then living. There was a "fly in amber" quality about much of late nineteenth- and early twentieth-century Newburyport, he felt; as if people preferred to cling to a past rosier than the present or future. Perhaps it was just that it was less expensive to maintain the old ways.

Whatever the reason, the old ways persisted in Newburyport and especially around Arrowhead. As people from more progressive places began to find this old-fashioned quality attractive, "summer visitors" began to arrive. As early as the 1840s and 50s, some of the Boston literati came to the area to visit John Greenleaf Whittier across the river in Amesbury; or Thomas Wentworth Higginson, who was briefly pastor of the First Religious Society in Newburyport; or Higginson's cousins, the Curzons, who lived at the old mill on the Artichoke. Later in the century, they came to visit Harriet Prescott Spofford at Deer Island. Some were so enchanted with the place that they returned to spend a few weeks in the summer, boarding with local residents.

One of the great attractions of the area was the natural stand of laurels along the bluff on the north side of Moulton Hill. The fame of "The Laurels" was enhanced by the Curzons' annual picnic there for their literary friends from Concord, Salem, and Boston, and during the short but spectacular laurel season flocks of nature lovers came to view them. The Bartletts, who owned the land, admitted them for a small fee.

Others found their way to the area, too. Uncle William's diary in August, 1858, speaks of "eight or nine fellows come round from Boston in a sloop boat, camped in Uncle John's pines." He rented his team to these men, and he and others after him occasionally carried groups of "summer people" to Plum Island or other picnic spots in their wagons. Thus did the hankering of the city folk after bucolic pleasures augment the economy of the farmers.

Summer people enjoyed jaunts to nearby beaches. LMP

Henry Moulton's castle on Moulton Hill, c.1900. LMP

In 1866, Captain Henry Moulton, a Civil War army officer and wealthy descendant of the Parsonsfield Moultons, purchased a piece of his "ancestral land" on Moulton Hill and at the crest of the hill built a replica of Naworth Castle, an ancient Moulton house in Warwickshire, England. Here he lived to the end of his life in 1896, steeped in the family genealogy he was preparing for publication.

Toward the end of the century, financier Edward Moseley chose the area at the curve of the Merrimack for his new country estate, "Maudesleigh." In the next two decades, "Lord" Moseley, as his plainer neighbors called him, bought up the farms surrounding Arrowhead. His son Charles also bought the old Bartlett farm near Bartlett's Cove and re-modelled it as his own country seat, which he called "Chailey." Gradually, the Moseley family converted some of the old houses in the area into care-taker's cottages and moved or tore down others—including Henry

Moulton's "castle"—to create the rolling fields and wooded bridle paths necessary for a proper English-style manor.

By the end of the nineteenth century, then, the neighborhood around Arrowhead had changed greatly. As in many other outlying parts of New England, much of the land had slipped out of intensive production and into a sleepy, picturesque, "rural" mode. Families moved away or dwindled to elderly spinster representation; houses and farm buildings disappeared.

The pressures on Arrowhead to do likewise in this period were great, and the acreage did in fact shrink. The old house built by the original William had vanished sometime before 1870; no one knows what happened to it. In the two eighteenth-century houses next to it lived Charlotte's grandfather Joseph (Joseph Bartlett Moulton, 1839–1915) and his family, which included several assorted aunts, uncles, sisters, and brothers— among them Rufus Wigglesworth, now retired from sea—as well as seven children. Five of Joseph's sons and daughters would remain on the farm to the ends of their lives; only one would have children.

While his cousins up the road were so numerous that the land was hard put to support them all, John Currier, an only child, had asthma and was not much suited to farming. Pressed for cash in the early 1890s, he sold a hundred acres to "Lord" Moseley. In 1897, the City of Newburyport, seeking to improve its public water supply, used its powers of eminent domain to acquire the land which sloped down toward the river. This lot, to which John Currier had moved his house in 1873, included much of the best pasture and crop land, and springs which provided most of the farm's water.

Henry Moulton's daughters on an afternoon drive to pick up the mail. LMP

GUERNSEYS, SUMMER BOARDERS,
AND FLOWERS

WITH the move in 1897 of John Currier's old house across the road to stand next to those of the rest of the family, Arrowhead as we know it today began to take shape. This household at that time consisted of John Currier and Orissa, their son Charlie, his maiden sister, Ida (Mary Ida Moulton, 1859–1934), sister Lizzie, and her husband, Charles Plummer. Charlie was then 21; he had for several years assumed more and more of the work of the farm from his father, whose health was failing, and from this time on his seems to have been the guiding hand.

Charlie took over the farm at a low ebb in its fortunes and invested the money from the forced sale of the hillside land in a fresh start—one designed to best utilize the land that was left and to take advantage of local and Boston markets. He extended the orchards; his plan, still in the slant-front desk, shows a number of new varieties of apples and peaches, plus some pears and cherries. These evidently did well, for we have a number of "First Premium" certificates from local fairs in these years, especially for peaches. He also planted raspberries, strawberries, Green Mountain potatoes, corn, peas, and other table crops for sale as well as for the family's use. The apples he shipped to Boston on the B and M railroad; other fruits and vegetables he sold locally.

High quality milk, which found a ready local market, was also part of his plan. The improvements made to the house and barn at the time of the move included a milk room and modern accommodations for the cows. Like many quality-conscious dairy farmers at the turn of the century, Charlie was a strong believer in Guernseys, famous for the richness of their milk. He gradually improved his small herd with them, winning prizes at the local fairs in the process.

Poultry rounded out the picture. Along the edge of the cow pasture, Charlie built extensive hen and turkey pens. Here he and Lizzie raised hens, and he briefly experimented with turkeys.

Up the road, John Currier's cousin Joseph's family carried on a similar mixed farming operation. Joseph's sons, Will (William Arthur Moulton, 1874–1945) and Sam (Samuel Cutting Moulton, 1884–1942) kept a good-sized herd of cows which they were upgrading with Ayrshire stock. They also raised large quantities of beans and potatoes and cut firewood and some lumber from their woodlots, while their sisters tended extensive home vegetable gardens.

This household included four daughters: Ruth (Ruth Page Moulton, 1872–1946), Nell (Ellen Bartlett Moulton, 1871–1941), Sarah (Sarah Lydia Moulton, 1884–1954) and Rebecca (Rebecca Jane Moulton, 1886–1967). The prettiest of these, Sarah, caught Charlie's eye. Courting her was not easy; Sarah's mother Ellen (Ellen Ordway Moulton, 1852–1931) seems to have taken the notion that Charlie should marry her youngest daughter, Rebecca—a notion that we're not sure Rebecca shared—and objected to the courtship. Like many others before and since, the young lovers persisted, however, signalling each other across the intervening field by raising and lowering their window shades. Perseverance paid, and in 1908, Charlie and Sarah were married. After a wedding trip to the White Mountains, Sarah moved across the field to her new home.

The household into which Sarah moved was, like the one she left, a comfortable Victorian extended family. It soon became predominantly female, and would remain that way for many years. Lizzie's husband, Charles Plummer, died in 1914; John Currier died in 1911. By the time Orissa died a few years later, Charlie and Sarah's two daughters, Elizabeth and Charlotte, had been born. Ida and Lizzie remained a vital part of the family, sharing the work of the house with Sarah. Ida did most of the cooking; Lizzie, laundry and sewing as well as tending the henyard and working in the kitchen garden; and Sarah kept the house tidied and polished and cared for the two girls.

By this time, Arrowhead had extensive ties with "downtown." Charlie regularly sold milk, eggs, fruits and vegetables to several small stores and private customers there, driving into town several times a week for this purpose. The family was active in the Belleville Church, rolling to church each Sunday in two buggies—the last family in the congregation to do so. Missionary Society and sewing for the Church Fair were important parts of social life, and the Belleville "book" club—an association for the exchange of magazines—provided a wide variety of reading material.

We have, fortunately, an extensive record of life at Arrowhead during the last years of the nineteenth century and the beginning of the twentieth in Aunt Lizzie's photographs. Working with glass plates that she developed herself—her cabinetful of chemical supplies is still in the attic—she recorded family gatherings and daily life on the farm. A few of her many pictures are reproduced in this book.

Lizzie's pictures show a life of hard work, cooperative effort and simple pleasures—evenings spent reading, playing checkers, making music; summer picnics at the beach; and much visiting back and forth with Charlie's sister Hannah (Hannah Moulton Merrill, 1864–1956) and her family in Hampton Falls. Outside, they show Charlie and others working in

*Charlie and Sarah on their wedding trip to the
White Mountains, 1908; and (below) with baby
Elizabeth, off on a Sunday afternoon drive.* LMP;
*(Opposite page) The Arrowhead photographer,
Aunt Lizzie, posing for her own camera.*

suit vest and trousers—a nineteenth-century mode of work dress he carried on to the end of his life, relegating his older suits to work status, and pulling on his "over-hauls" over them when he worked in the barn. Inside, they show a comfortable melange of factory-made Victorian furniture and older pieces set against the light, airy, uncurtained windows popularized by the Beecher sisters' homemaking books, and complemented by a luxuriant array of house plants.

Lizzie's interest in photography seems to have waned somewhat after 1910, but in this later period we have a charming series of albums taken by Grace Collins, a summer boarder. We don't know exactly how early Arrowhead began to take in boarders, but the practice continued until about 1930. "Regulars"—who in the Teens and Twenties included an elderly bachelor who lived nearby, a brother and sister and their aged mother

from Newburyport, and the aforementioned Grace Collins, a schoolteacher from Virginia—became a sort of extension of the family. Grace Collins' idyllic pictures of berry-picking expeditions and leisurely afternoons on a sun-dappled lawn, together with the well-worn recipes for delicious dishes in Aunt Ida's cookbook, suggest why they came.

Compared with Aunt Lizzie's earliest photographs, these pictures show an interesting change in the landscape around Arrowhead. The earlier views show broad, open fields and neat but spare yards; in the later ones, trees and bushes line the stone walls and creep in at the edges of the fields. These differences reflect the shifting patterns of life in the area: less intensive farming and more enjoyment of the "picturesque." Once the constant plowing and mowing ceased, how quickly the trees reclaimed the land so arduously cleared two centuries before!

*Two of the Floyd sisters, and
(below) John Currier, son Charlie,
and a Merrill nephew play cards.* LMP

*Charlie plays his guitar while
Aunt Abby Bartlett holds her cat,
Tombo Sunshine, and Ida reads.* LMP

*John Currier heads to town in the Democrat wagon (above);
and the tin peddlar stops in front of the house. At right,
visiting Floyd cousins (above); Lizzie and Charlie play checkers;
Lizzie and her husband, Charles relax in their bedroom.* LMP

Hampton Falls cousins wading;
picnics at the beach or by the
side of the road (lower right)
were favorite pastimes. LMP

 *Will brings hay along Ferry Road while Charlie and the girls set out to pick up a new load (below). Opposite: the Doll Table Committee prepares for a church fair; Elizabeth and Charlotte share the swing; Charlotte, with pony and Duke.* LMP

"A passer-by on Ferry Road? Must be lost!"

Around the houses, too, the greenery flourished; in this case the trees and shrubbery that softened the lawns resulted from greater interest in — and leisure time for — gardening. Rebecca planted perennial borders around the kitchen garden, and shrubs near the house. Ida carefully nurtured the old rosebushes, bridal wreath, mock orange and lilacs transplanted from across the road with the house, and added hydrangeas, wygelia and forsythia, while Charlie planted maple and birch trees. Sarah raised sweet peas by the back door.

Enchanted by the gardens at Chailey, Elizabeth at nine longed for one of her own, complete with little beds and paths; and Charlie helped her lay one out near the workshop. Small at first, it grew as she proved she could take care of it. A father-daughter excursion to the pasture yielded red cedar posts and branches to be fashioned into rustic rose arbors and trellises. A few years later, she added a bed of asters which she marketed to a local florist as her first cash crop.

Like their counterparts in earlier generations, Elizabeth and Charlotte were expected to share in the work of the farm and household. Elizabeth kept the extensive lawns mowed and trimmed around the shrubs, not a small undertaking with a hand mower, and she also had the responsibility of arranging all the flowers for the house. Charlotte helped Charlie in the fields by riding the horse when he cultivated and furrowed. Both girls helped pick the many boxes of strawberries to be sold in town, as well

as those for home use. But life on the farm wasn't all work. The girls had
a pony which Charlotte, especially, enjoyed riding with Duke, the Scottish sheep dog, trotting alongside. In the summer evenings there were
long croquet games on the side lawn, and in winter Moulton Hill furnished
a natural ski slope.

The twenties didn't exactly roar at Arrowhead, but they did bring
changes. Carrying the tradition of the farmer as mechanic and craftsman
into the twentieth century, Charlie installed the first indoor bathroom in
1925, and in 1927 he wired the house for electricity. The latter innovation
paved the way for such improvements as a new copper-tubbed washing
machine with an automatic wringer, electric irons and toaster and a radio
to accompany the evening pastimes of sewing, reading, and parlor games.
Gradually, similar changes worked their way up the road to the other two
houses.

Supplemented by the summer boarders and the sale of a little gravel
from a corner of the cow pasture, the farm which Charlie was working
by himself with occasional hired help was providing a comfortable living.
In the early Twenties Sarah even had a new fur coat, and in 1922 Charlie
bought his first "machine," a Dodge touring car, the first automobile in
the immediate neighborhood, and converted the workshop into a garage.
But it would be several years before Ferry Road in the spring could sup-

Charlie's first "machine," a 1922 Dodge.

port such modern transportation, and the horse and buggy often headed off through the mud for town while the "machine" stayed snug in its garage.

Other things changed more slowly. Aunt Ida continued to cook on a combination coal and wood range, although it was a new one with a thermometer on the oven door, and Sarah still made her famous ice cream

Aunt Ida's new cookstove, 1930

in a hand-cranked freezer. Most of the food still came from the farm: fruits and vegetables, fresh in season, "put up" and stored in neat rows of glass jars on shelves in the cellar for winter, beans grown on the place and dried for baking, eggs and Sunday-dinner roast chicken from the hen pen, and rich Guernsey milk and cream from the barn. In addition to supplying milk for sale to a local dairy, the cows at Arrowhead long protected the family from the detested "burned taste" of the pasteurized milk required by twentieth century health laws.

"Store-bought" foods had also changed little. Butter, along with peanut butter and cheese, came from the "butter and egg" store downtown, where the proprietor unerringly scooped or cut the exact amount requested from a huge tub. Molasses still came from the big barrel in "Tumps" Bowlen's store on Merrimack Street, carefully funneled into the family molasses jug.

1930's electric toaster

Flour and sugar came in barrels, too; each fall Charlie laid in a barrel of bread flour and a barrel of pastry flour, which he paid for with apples and potatoes.

A frequent visitor to Arrowhead, in 1929, Kendall Blanchard (who liked it well enough to become part of the family by marrying Elizabeth), recorded his impressions:

Life at the farm is as beautiful and simple as the surrounding country. How snug and happy these farm folk are when they gather around in the living room after the day's hard work. The room has antique furnishings and is lighted by an oil lamp which gives a soft warm light. Many evenings I have sat with them and held pleasant conversations. The children would indulge in some interesting reading in a corner; the elderly people would doze off for a nap, and some of us would remain talking for a long while.

As the industrialized society suffered through its most severe depression, the national hard times of the Thirties had little visible effect on Arrowhead. Cash was, of course, tighter. But Newburyport's economy had

Family's molasses jug

already been depressed, so the Great Depression brought less dramatic change here. And the style of life at Arrowhead—food grown and prepared at home, clothes largely made at home, too, an attic full of whatever might be needed in household furnishings, simple tastes and old-fashioned pleasures—was well designed to weather such an economic storm.

But times were changing. Charlie, Sam, and Will were all growing older; there were no sons to take over the work; and there was no money to invest in modern machinery. The girls, Elizabeth and Charlotte, grew up and married. Elizabeth and Kendall moved away permanently, but after a year of living in Belleville, Charlotte and her husband, Glen (Glendon Woods Chase, 1909–1981) moved back to the farm in 1938. Glen had grown up on a farm in Newbury, gone to "Essex Aggie," and worked in his father's dairy. He had first come to Arrowhead to buy sweet peas for his mother and grandmother—sweet peas that Charlotte had added to the garden she had taken over from Elizabeth. Now he continued to drive a milk truck for a local dairy while he and Charlotte built Sunnyslope Greenhouses.

Beginning with annuals started indoors in the spring, which Charlotte had helped her Aunt Lizzie do for the family garden as long as she could remember, they opened a small stand on Glen's family land in Newbury. Gradually, in the 1940s and 50s this grew to include several greenhouses, an extensive retail business in cut flowers and plants of all kinds, and wholesale growing of poinsettias, cyclamen, Easter lilies and geraniums.

Charlotte and Glen on their wedding trip. CMC/GWC

Charlotte, Charlie, and companion in the best part of haying.

Glen had an uncanny knack for producing luxuriant, healthy plants, Charlotte had a talent for floral design, and both had an avid interest in unusual species and new varieties. Both would eventually win awards from the Massachusetts Horticultural Society—Glen for excellence in growing, Charlotte for design. With emphasis on quality and diversity, the greenhouses flourished, and Glen eventually left the dairy to concentrate entirely on horticulture.

Glen and Polly escort Arrowhead ducks to the pasture pond. CMC

Dick took to farming early and naturally. He enjoyed the fruits of his first corn crop, and learned to handle the tractor. CMC

As the greenhouses grew, farming at Arrowhead diminished. Failing health forced both Will and Charlie to sell most of their cows and cut back on crops. In the early 1940s Will sold his woodlot, and soon after that both he and Sam died. Rebecca, who inherited their land, kept extensive gardens, continued her long-time business of boarding pets, and sold hay in the fields, but farming virtually ceased on this part of Arrowhead by the mid-1940s. Charlie kept two or three cows and some hens and raised some vegetables for sale into the early 1950s. After Sarah's death in 1954, he, too, gave up most commercial operations, selling only eggs, hay in the field, and increasing amounts of gravel.

The farm's acreage continued to dwindle. Rebecca, feeling that nobody in the family would ever want to farm again, first leased and then in 1951 sold twenty acres of rich land sloping down to the river. In 1953, the city again exercised its power of eminent domain and took part of the cow pasture for the city water supply. This lowered the water table significantly, and dried up the pond which had always supplied water for the cows—and for Charlie's cherished collection of pink and white waterlilies. At about the same time, the Commonwealth of Massachusetts took a tiny corner of the same pasture for a new highway, I-95, then being built from Boston to New Hampshire. Charlie sold extensive amounts of gravel from the pasture for that highway.

A century after Ferry Road had been incorporated into the city of Newburyport, urbanism was beginning to encroach on the immediate neighborhood of Arrowhead. In the 1940s Ferry Road was oiled all the way up to the old house on the brow of the hill. Moreover, that house was now known as "Number 133 Ferry Road." Mail no longer came addressed to "the yellow house next to William Moulton," but to "129 Ferry Road." After the war, Will's woodlot became "Sandy Acres," and burgeoned with tiny, boxlike houses. And, to add insult to injury, the fields along Ferry Road began to be taxed as house lots!

As farming at Arrowhead decreased and urban influence increased, the pattern of life inevitably changed. The family living on the place in the 1940s and 50s consisted of Charlie and Sarah, Charlotte and Glen and their children Polly and Dick, and, up the road, Rebecca and her husband David Duffie, a retired Canadian horse trainer whom she had married after her first husband died.

Rather than supporting the family on it, the farm depended on outside income to support it, especially as real estate taxes rose. Except for summer vegetables, a little fruit, and some milk and eggs, most of the food now came from the store. Cooking however, changed very little, although Glen and Charlotte had introduced more modern appliances. What came home from the store were the raw materials for the time-honored favorites; Arrowhead had little use for such modern fare as "bought" cookies, the new cake mixes, canned soups (which Glen christened "bellywash"), or for that matter for the exotic casseroles, canapés, and other concoctions then being featured in the women's magazines which came from the Belleville book club.

Still, family tastes had changed somewhat over the years. During World War II, when Sarah patriotically tried replacing sugar with molasses in apple pie, as Charlie remembered his mother's doing it, the result was not even popular with Charlie. Nor was Sarah's wartime experiment with making soap as her mother had done any more acceptable; the hard yellow cakes languished on a shelf for years afterward. Butter was another matter, however; letting the cream rise and getting out the old churn was vastly preferable to having anything to do with oleo. Another wartime project, fattening a steer for family consumption, was also popular, but not repeated.

Old habits died hard, and despite the "creeping urbanism" and the decline of commercial farming, Arrowhead retained a farm character. The buildings and the land were there, and the family enjoyed using them. Even after giving up sales, Charlie and then Charlotte maintained enough hens to supply eggs for the family. Rebecca kept goats for several years. And Glen, Polly, and Dick particularly liked experimenting with various types of livestock; after the steer came two pigs—Susie and Sadie the Lady—a dozen ducks, some geese, and an assortment of bantams; plus a more standard array of pet dogs, cats, turtles, and guinea pigs. And the garden never stopped supplying all the vegetables and fruits for the table. To the traditional crops at Arrowhead, Glen added new varieties of raspberries and blueberries, plus the luxury of tomatoes and lettuce from the greenhouse year round.

The kitchen remained very much a farm kitchen, too; in fact it was still almost as much the center of family activity as the first William's kitchen had been. A sprawling room with eight doors leading to other parts of the house and large windows looking out over the fields, it contained a comfortable clutter of old tools and furniture as well as modern appliances. Aunt Ida's old range never left when the new gas range arrived, and a fire in it often took the chill off an October morning. In the center, the kitchen table still held sway, surrounded by Hitchcock "fancy chairs" Great-Grandmother Bartlett had bought. On this table Sarah cut out dresses, Polly did schoolwork, Charlotte put up jelly, and Charlie repaired the wooden works of the Grandfather clock whenever necessary. Around it, at various times, winter carpentry projects took shape; wallpapering progressed; Polly and her Portland cousin, Polly Blanchard, roller-skated during Christmas vacation; and Dick kept turtles and polly-wogs.

Flowers, not surprisingly, became a more important part of life at Arrowhead in these years. Glen's part-time job of managing the

The kitchen sink with orchids in bloom

Maudesleigh greenhouses provided both income and pleasure. Interesting new treasures moved back and forth between Maudesleigh, Sunnyslope, and Arrowhead, and many a rare cutting "fell into Glen's pant cuff" as he walked in another greenhouse or garden and came to thrive in a hospitable spot at Arrowhead. Thus, the garden which Elizabeth had started grew and evolved, reflecting Glen's and Charlotte's interest in unusual varieties, both new and old, and in cultivating wildflowers. Agapanthus, tuberous begonias, giant hybrid clematis, French lilacs joined the traditional shrubs and perennials, and wild meadow rue, lady slippers, trillium, and sweet fern nestled in sheltered corners. Gradually the garden assumed the character it still has today: groupings of semi-naturalized plants accented in summer with greenhouse exotics, a charming informal spot filled with pleasant memories. The house, too, overflowed with flowers—freesias, ranunculas, camelias, malmaisons, sweet peas, according to the season—and with plants—cyclamen, azaleas, gloxinia, clivia, orchids, orange trees.

Gardening, both flower and vegetable, remained very much a family affair at Arrowhead. Polly grew her first prize-winning zinnias at seven (with a little help from Glen); a few years later, under duress and to add to her country and western record collection, she picked raspberries at 7¢ a box. Dick, from the very beginning, was more interested in vegetables.

By the time he was eleven he was pulling a red wagon of corn, grown
on the farm, down the road to the suburban houses at Sandy Acres and
selling it door to door.

OLD LAND, NEW WAYS

YOU can lead a horse to water, the old saying goes, but you can't
make him drink. Polly's initial skirmishes in the raspberry patch with
agriculture as a source of income convinced her that greener fields lay
in the city, and she has been trying ever since to convince the family that
moving to Boston from Essex County is also a time-honored Yankee tradition. Dick's first ventures, on the other hand, soon blossomed into a
career. At thirteen he planted, with the help of a local farmer who was
then leasing land at Arrowhead, one and a half acres of corn; to sell it,
he set up a stand on Sunnyslope land on the High Road in Newbury. In
the next two years, with the aid of his trusty Gravely rototiller, he added
a quarter of an acre of other vegetables and expanded to three acres of
corn. By the end of his high school years, he was growing four or five
acres of corn, one and a half acres of potatoes and two acres of other
vegetables.

   Free at last from the demands of school in 1970, Dick bought a tractor
and escalated his operation: thirty acres of corn, twenty acres of vine
crops—squash, pumpkins, melons—and fifteen to twenty acres of
vegetables. In 1971, he built a plastic greenhouse on the farm and began
raising hothouse tomatoes. By this time, he was wholesaling to the Boston
market and was beginning to lease land in Amesbury, Salisbury, West
Newbury—wherever he could find it—in addition to the home place. Not
only did he need more land, he needed land that could support corn. After
successive water takings by the city, Arrowhead's natural water supply
had been diminished to the point where little of the land could be used
for corn, and the city main serving the end of Ferry Road was insufficient
to provide irrigation. In the late Seventies, he added seventy-five acres
of ensilage corn and 175 acres of hay and alfalfa.

   In these years, Dick was learning to farm in the time-honored—and
most effective—way: through his own experience combined with that of
others. He had already absorbed much farm lore from Charlie, and he continued to benefit from Glen's advice on growing. He also picked up ideas
about both the growing and the business ends of farming from other older
farmers in the area—Bill Pettingill in Salisbury, Albert Elwell in West New-

bury, Calvin Learoyd in Rowley—and from visiting other farms throughout New England, New York and Pennsylvania. While some professionals jealously guard their secrets, farmers are almost always willing to share theirs, and to benefit from each other's mistakes and successes, so a beginner can get a big leg up in this way. Together with the land grant college extension services, which function as a link between private and government agricultural research and commercial growing operations, this informal information network among farmers is indispensable. A pool of shared information, however, must be coupled with individual experience; and, of course, it has to be continually modified as economic conditions and cultural practices change.

Throughout the Sixties and Seventies, then, Dick was constantly experimenting with new crops and approaches, learning what worked and what didn't, and what he did and didn't like to do. Livestock, for example, is always an attractive option for farmers, and he made several ventures in this direction. Like Charlie before him, he experimented, briefly, with turkeys; like Glen, he tried his hand at pigs. He also tried raising a few heifers before concluding that livestock was not for Arrowhead.

Meanwhile, the Arrowhead Farm stand was growing and becoming an established part of the scene on High Road. Its market was growing,

Trying out the new Gravely in the mid-1960's. CMC

too; a renaissance was taking place in Newburyport. After nearly a century as a backwater, the old port city had become a part of the ever-widening exurbia around Boston. Drawn by its handsome old houses and restored Federal-period downtown, a young, affluent group of people with sophisticated tastes was moving in to refurbish the old homes and put down roots in this charming old city. Fresh, locally grown produce appealed to them.

The Arab oil embargo of 1973 and its aftermath brought a dramatic shift in the economics of farming. As petroleum prices skyrocketed, farm production costs soared, but wholesale prices for farm goods did not. From twenty-five percent of the total cost at wholesale in the 60s and early 70s, direct production costs before harvesting and marketing rose to fifty percent by the 1980s. "Grow everything you sell" had once been the maxim for farm stands. Then farmers planned to grow more crops than they needed, to guarantee they would have enough to meet demand; if they had too much, they could wholesale the surplus or leave it in the field. Now, the precept became "sell everything you grow," and fill in the gaps by buying. Other things changed, too. By 1978, the cost of oil had made hothouse tomatoes too expensive to grow in Massachusetts: no one would pay for them.

Oil costs were turning the greenhouse business topsy-turvy in the 1970s, too. Cut flowers could be shipped by air from South America, Korea, and New Zealand more cheaply than they could be raised in Massachusetts

The Arrowhead Farm Stand in the late 1960's. CMC

greenhouses, and greenhouse customers were not willing to pay the runaway costs of production for specimen cyclamen and Easter lilies. They were more interested than ever, though, in annuals, perennials and hanging plants for the gardens and porches of their restored houses. Farm stand customers were also increasingly interested in specialty items: herbs, dried flowers, plants, cut flowers and Christmas greens.

And so an amalgamation was born. By closing the Sunnyslope greenhouses during the coldest months and starting them up in March, plants could be produced without exorbitant cost and sold efficiently through the farm stand. This also extended the farm stand season: open at Easter with Easter plants; stay open through Thanksgiving and Christmas with "ornamentals" and Christmas greens. The greenhouses could also continue to start the field transplants necessary for early crops to "beat the season" and bring the best prices.

By 1981, Dick had virtually eliminated wholesale, and cut his acreage from 360 to 65, with plans to cut 20 more and add irrigation. The final cut would limit his entire acreage to the tillable land on the home place, plus the twenty acres across the road by the river, which Rebecca had sold to Bill Pettingill, and which Bill now leased back to Dick. The future of the river field as farmland has recently been assured by the sale of development rights to the Commonwealth of Massachusetts.

Isn't such a small farm old-fashioned? Well, yes, in some ways. It means much more hand work, because specialized machines, such as mechanical pickers, are not cost-efficient on small acreages. Whatever machinery is practical, however, is used; a farm of this size will usually have as much as $100,000 invested in machinery.

A small farm also requires intensive use of the land; finding productive uses for odd corners of the land which farmers would have ignored a few years ago—uses much like those earlier generations found for the same land. Arrowhead today still produces much of the lumber for its buildings and fences, and a couple of years ago, Dick embarked on a tree-planting program. Each year he plants an acre of trees on bits and pieces of land unsuited to other crops—Scotch pine on high, hard-to-till slopes, fir and spruce on low land too wet to plow. At the end of eight or nine years, he will harvest an acre a year, or about 1,200 trees, to sell as Christmas trees at the stand, and replant in the spring. A side benefit of this program is that planting all areas of the farm, by discouraging growth of brush, cuts down the opportunities for the development of "woodchuck condos" in these areas, thus reducing a serious nuisance to market gardening.

Careful planning and double-cropping of early and late crops also maximize the use of the land and cut down on late season weed growth. Spring crops such as peas, greens and strawberries are followed with later crops

Plants and fresh flowers from the Sunnyslope
Greenhouses augment the vegetables at the stand
(below). Charlotte arranges a hanging plant. CDC

Dick examines local apples with Bill Pettingill, an area
farmer from whom he has learned much. PCH.
Christmas greens and trees extend the stand's season (below). CDC

like summer squashes, zucchini, beans, cucumbers and the cole crops—cabbages, broccoli, cauliflower, brussel sprouts, turnips, radishes, kale, and collards. Planting cover crops—rye in winter, buckwheat, Sudan grass or oats in summer—and plowing them in builds the soil.

In other ways, the farm is entirely modern. Petrochemicals and commercial fertilizers, although expensive, are cheaper than the transportation and handling of cow manure, so Arrowhead uses them. To these we may soon be adding composted city sludge. As new equipment is needed, gas-powered machinery is replaced with diesel. New varieties with superior characteristics are constantly being tried. At present, Arrowhead is cooperating with the University of Massachusetts in testing varieties of asparagus for resistance to fusarium root-rot. Mechanical scrubbers clean gourds and other produce. Cloches, or row covers, used successfully for several years in Japan and western Europe, are now being introduced to increase yield and lengthen the season.

Although help for processing and marketing crops—machine operators, farm stand clerks and greenhouse workers—are not hard to find, people to do hot, dirty, hand field work are very hard to find. Machines and chemicals that will take care of these jobs are indispensable. Black plastic mulches and specialized herbicides, both applied by machine, cut down significantly on hand cultivating and weeding, and Arrowhead had the first plastic mulch laying machine in the area. Pick-your-own and smaller volume retailing instead of wholesaling reduce the need for hand harvest workers.

Modern marketing techniques at the farm stand round out the picture. Promotions include fruit samples, take-home samples of new vegetable varieties, recipes, newsletters to customers, and stands at fairs and farmer's markets. The emphasis here is on top quality, friendly service and education of customers. The staff offers advice on preparation of fruits and vegetables, and on gardening. We also try to offer new varieties of old favorites, and to introduce our customers to unusual garden plants and forgotten old-fashioned varieties.

The stand, in fact, is the key factor in shaping Arrowhead operations today. Almost all cropping is geared to sales at the stand, which is open seven days a week from Easter until Christmas. Easter plants and over two hundred varieties of bedding plants form the major spring crop. Ornamentals are in general more profitable than vegetables and enjoy good sales in all seasons. After bedding plants come cut flowers in summer, gourds and dried flowers in fall and, finally, Christmas greens. Christmas sales are growing every year. The farm woodlot supplies cedar, juniper, white pine, laurel and other greens which are turned into old-fashioned quality wreaths, garlands and other decorations.

*A field of mixed vegetables (above);
Ken Groder, Glen, and Dick comparing notes
at the Sandwich Fair, 1968.* CMC

A new tractor: no more down time, but how will we make the payments? CDC

Vegetables are still the heart of the operation in summer, however. Major crops, each of which grosses in the tens of thousands of dollars in a typical year, include twenty-five acres of sweet corn, four acres of berries, two acres of tomatoes, three acres of salad crops, ten acres of vine crops, plus an acre each of cut flowers and dried flowers. A number of smaller crops, including various kinds of beans, melons, root crops and asparagus, gross a few thousand dollars each year. Many others, such as herbs, kale, collards and Chinese vegetables, gross much less—some only $10–$50 a year—but are grown because they attract customers to the stand.

A few seasonal crops are grown for wholesale because, while they are not very profitable, they bring in cash at times when stand sales are slim. These include greenhouse crops such as herbs and other four-inch pot plants sold to chain stores in March and April; and vine crops and dried flowers sold through brokers when stand sales drop off after Labor Day. Evening out the cash flow in this way reduces the perennial need for borrowed money.

The exact mix of crops and the marketing techniques continue to shift slightly each year, of course, as market conditions and other elements of the complex farming equation change. Current plans call for further expansion of retail sales. The roadside stand selling season continues to lengthen, with fall sales of gourds, dried flowers, Indian corn, cider, apples and fall vegetables. With better storage and sales facilities, year-round operation might be possible. Together with mail-order sale of specialty crops, this would further reduce dependence on wholesale and improve both overall cash flow and profitability of these crops. Even more diversification of crop mix and better irrigation to insure production of quality crops are also planned.

Some version of this smaller, more intensive and more diversified farm is probably the wave of the future in such densely settled parts of New England as eastern Massachusetts. Farmland here is rapidly disappearing. Arrowhead is the only active market garden left in Newburyport, and much of the land Dick was leasing in the mid-70s in nearby towns has become houselots. As leasable land decreases, and as the cost of gasoline to transport equipment and help back and forth increases, more and more farmers are looking at the option of cutting back on acreage. "The farmer's footprint is the best fertilizer," says an old adage; it may prove as true at the end of the twentieth century as in earlier generations.

Like farming, life at Arrowhead today combines the old and the new. Justin still enjoys grinding coffee in the old coffee mill, but most of the time our coffee comes from the electric grinder via the coffee machine, and for iced coffee, there's a wonderful little cube dispenser in the refrigerator door. On the other hand, the wood stove is a mainstay for heat again, with the oil burner viewed strictly as supplement. (And we never

The produce scrubber at work. CDC

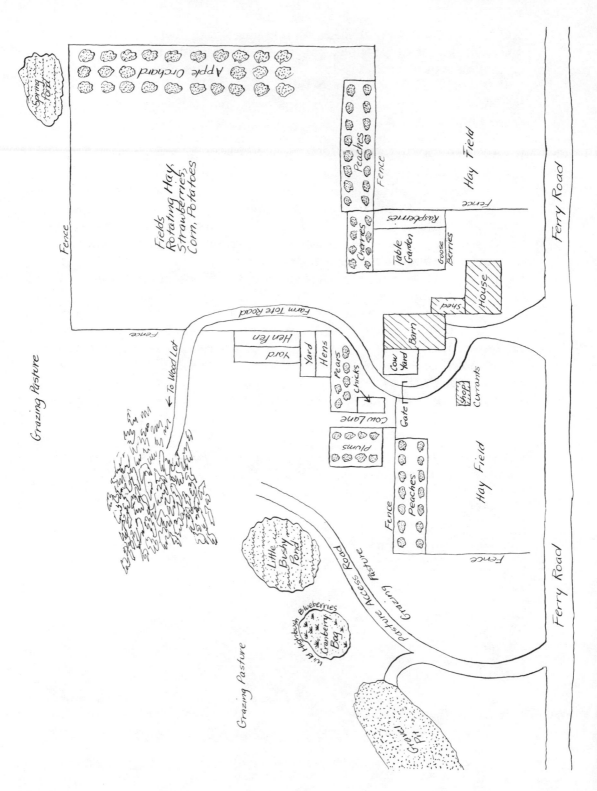

did give up the old custom of closing off rooms in the winter!) Much of this book has been written around the kitchen table with a cozy fire in the wood stove and an iron teakettle gently hissing. And as we tested the recipes, we carefully saved, as we've always done, the "orts" for the hens.

Arrowhead is still Arrowhead. The acreage has grown and then diminished over the centuries; new machines and new methods have replaced the old. Trucks and vans bounce up and down Ferry Road, supplying the busy farm stand at the other end of town. But the land is still the same, and it is still spacious and airy.

In the spring, the pink lady slippers still bloom on the steep slopes of the kettles in the old pasture, and the pine trees murmur together above them in the soft south breeze.

In the summer, the river winds still cool the piazza where we shell peas in the hammock. As the day grows hot, the blue haze gathers against the shape of Moulton Hill, and the scent of warm strawberries wafts across the field.

In the fall, red maples, yellow birches, and ruddy brown oaks still rim the bronzy fields for a while before their leaves are swept into nooks and crannies to blanket and help renew the land.

In winter, the wind still whistles down the Merrimack, and those of us who live in the northwest bedrooms still have to be hardy. It drives swirling waves of snow, cold but beautiful, before it, and sometimes the old houses shudder with its impact. But they still stand.

"Keep the land, and the land will keep you." We like to think that the first William, with his fulling mill and his store, would understand the Christmas trees, the dried flowers, the farm stand with its pumpkins sold more for decoration than for sustenance. That Great-Grandmother Bartlett wouldn't mind her "fancy chairs" being used in the kitchen. That Uncle William, who cut sticks for the shipyard, would approve of the new barn built with lumber from the same stand of trees. They understood survival in New England, after all.

Arrowhead bows to the past, and moves on into its fourth century.

Arrowhead Through the Seasons

*To every thing, there is a season, and a time to every purpose
under the heaven . . .*

ECCLESIASTES III:1

THE SEASONS AT ARROWHEAD

IT **IS** the natural condition of mankind that the pattern of life varies
greatly with the seasons, especially in the temperate zone, where the
climate differs widely from summer to winter. In the past century, the
Industrial Revolution and twentieth century technology have altered this
ancient fact. Central heat and air conditioning, paved highways and motor
vehicles, foods conveyed by air from other climates overnight, jobs in factories and offices have smoothed the effects of the seasons to barely a ripple.
Who today remembers the "six weeks' want," when the root cellar had
been exhausted and nothing green graced the landscape? Or the stern real-
ity of "February second is Candlemas Day; you should have half your wood
and half your hay."

Traditional New England farm life is still governed by the seasons.
For three hundred years at Arrowhead April has meant planting peas;
June, searching the sky for signs of rain and hurrying to get the hay in;
September, being on alert for the first frost. And while we, too, now have
central heat, mud-free roads, and access to the convenience store, life at
Arrowhead is tied to the seasons.

Farm people are moved by the seasons; the spring sun and the southwest wind stir them as surely as they stir the plants to begin the cycle
again. There is a special feeling that comes from making things grow, from
working with the land; and land, weather and seasons are inextricably
interwoven. Each farm has its unique combination of weather and land,
what will work on one place won't work a mile away. Working the same
land year after year, you get an understanding for its peculiar personal-
ity, its idiosyncracies, what it will and won't do in any season. This may

explain why farmers are often said to have an "innate feeling for the land." While this understanding may not be born into the child, it is certainly bred into him: from earliest memory the farm child hears about the seasons, the weather, this field or that field, until it is second nature. No amount of education can replace this lifelong interaction.

Dick has a theory that farmers are "solartics" (not to be confused with lunatics, whose madness is derived from the moon), whose reason is affected by the sun. As the sun slips away to the south and summer eases into fall, farmers become increasingly tired and discouraged. The long season of too much work, too many hassles and growing and marketing conditions beyond their control has taken its toll. By October, they begin to feel farming is hopeless. By November, older farmers have decided that this is the year to retire; younger ones are planning to cut back on acreage. But as surely as the February sun moving northward brings a faint new warmth, it also brings a new optimism. "Maybe just a few acres, for one more year," say the old ones. Maybe a few more acres of corn," say the young, "and a few of zucchini, and . . ." By March, when the sun turns the frozen earth to mud, they are tramping the fields, assessing how soon they can get a tractor on them, and by April, they have sent in an order for additional seed.

Once the cycle has begun, tasks on the farm move in a predictable pattern dictated by the mix of crops and the progression of the seasons. Dates vary from year to year, depending on variations in sun, rain and temperature, but the rotation of planting, tending and harvesting governs the day to day life on the farm.

If the activities and attitudes on the farm differ with the seasons, so, traditionally, has the food. We still associate fresh fruits and vegetables with summer, applesauce and apple pie with fall, and Indian pudding with winter. For earlier generations, seasonal associations were far more profound: you ate what you had, and you had different foods at different times of year. Milk and eggs, for example, were plentiful in spring and summer, scarce or nonexistent at other seasons. Winter fruits and vegetables were limited to those which could be stored in some way or dried. Farmers butchered livestock in the fall to reduce the amount of feed needed for the winter; fresh meat was plentiful at slaughter time, salted and smoked thereafter. Fresh, tender lamb came in the spring, three months or so after the ewes had lambed.

Traditional New England cooking, like husbandry, then, has an inevitable seasonal cycle.

We are less bound by it today, of course, and many foods taste good at any season. At Arrowhead we often find, however, that foods taste best at their appointed season, and we are most likely to serve them then. For that reason, we have organized the cookery section of this book by season.

". . . if I could sit down to dinner on a piece of their excellent salt pork and pumpkin, I would not give a farthing for all the luxuries of Paris."
BENJAMIN FRANKLIN

FOOD ON THE FARM

EVEN among the legions of transplanted New Englanders who have waxed eloquent with nostalgia for the food of their youth, "plain, honest fare" has always been the most popular phrase to describe New England farm cooking. Poorly executed—over-baked, over-boiled, watered-down, or mixed with a leaden hand—Yankee cooking can be just plain boring. But well-prepared, carefully timed, with utterly fresh top-quality ingredients, their natural flavors enhanced by the most sparing hand with condiments—it is the stuff of fond memory.

"Plain, honest fare" meant—and still means—simple, straightforward preparation of high quality, nutritional foods. Meats, roasted or pan fried; and vegetables, boiled or occasionally baked, stand on the merits of their own good flavor, unaltered with marinades, ungussied with sauces and garnishes. For this reason our cooking section, like the earliest New England cookbooks, may seem short on recipes for meat and vegetables. This brevity reflects the simplicity of their preparation rather than the extent of their representation at table.

While many have been grown from the earliest days in New England gardens, herbs have not figured largely in New England cooking. Sage and thyme went primarily into making sausage, or into stuffing for poultry; peppermint or spearmint flavored candy or tea, or soothed an upset stomach. Indeed, most herbs in early New England gardens were grown for their medicinal, rather than their culinary value. In recent years, however, we've enjoyed growing herbs in the garden and also in the kitchen, and some of the recipes we've included here, such as Tarragon Chicken and Herbed Pork Chops reflect these additions.

Spices are another story, however. The earliest Yankee ships carrying salt fish and barrel staves to the West Indies brought back pepper, ginger, and allspice, as well as molasses and rum, and these became basic parts of the New England cuisine. Before 1800, trade with the Far East added cinnamon, clove, and nutmeg to the repertoire. These combined particularly fortunately with the regional staples of pumpkin, squash, apples, and molasses, producing some of the most characteristic flavors of Yankee cooking. Other condiments were available, too. Mustard, horseradish, and celery seed, if not native, were growing nicely in New England soon after settlement, and cider vinegar soon became a staple, as well. Perhaps because many of them were expensive, or perhaps simply out of preference, New England cooks have used these flavorings sparingly.

By today's standards, the traditional New England diet seems heavy and rich, full of both calories and cholesterol. Meat and potatoes (with gravy, of course); vegetables heavily tending to corn, beans, and peas (with plenty of salt and melting butter); fruits made into pies and puddings or doused with thick cream; and plenty of breads and cakes—how could such calorific fare have produced the gaunt Yankee farmer of stereotype? One explanation is that the farm people who ate this food—men, women and children—did hard physical work all day, every day. They also lived in houses that were very cold in the winter. The daily pattern of Yankee farm life quickly converted fat into muscle power and body heat.

And, if we look closely at how farm families actually ate, the picture changes somewhat. Many of the richest dishes were once-in-a-while treats; others were strictly seasonal. Meat was not part of every meal; often it was served more as a condiment than a main part of the menu: a bit of bacon or salt pork in the baked beans or the corn chowder, a little sausage with baked potato and applesauce. Moreover, meat was less fatty (albeit less tender) before modern breeding and grain feeding were introduced. Vegetables and whole grains were mainstays, and many traditional New England recipes, such as succotash or corn chowder, feature combinations of vegetables which provide complete protein.

It is a rich diet, however, and in recent years we've made some concessions and modifications to it at Arrowhead, or we'd all be quite roly-poly. Smaller portions, low-fat milk, fewer pies and cakes, and cream and butter as strictly occasional treats are part of the realities of a more mechanized life. Allergies, and low-salt and low-cholesterol diets have also caused us to modify some of our favorite recipes so that everyone in the family can enjoy them. Such considerations seem to be widespread today, so we've included some of our more successful substitutions in the recipes which follow. What we have *not* altered is the insistence on high quality and absolute freshness in all the ingredients.

Traditionally on the farm breakfast was a hearty meal, geared to fill appetites whetted by barn chores, starting the fire and drawing water, and other work already done, and to fuel the morning's work. It consisted of porridges of corn, oats or other grains, served with butter, milk, or cream when they were available; fried puddings, johnnycake or hotcakes with butter, molasses, or maple syrup; "breakfast gems" (muffins) or other breads with butter and perhaps jam. Coffee was customary from the eighteenth century onward; eggs or meat might or might not be added. A generous slab of pie, preferably apple or squash, was often included.

Dinner, the heavy meal of the day, was served at noon to sustain the afternoon's work. On a prosperous farm the main dish was usually meat or poultry, but a hearty chowder or stew might substitute. Potato and another vegetable, bread and butter, and dessert rounded out the meal.

Supper was a light meal to end the day, often consisting of little more than bread and butter served with fresh or preserved fruit, milk and tea, and dessert. Crackertoast—oven-crisped pilot crackers broken up and served with butter and hot milk—asparagus on toast, or bread and butter with rhubarb or applesauce, made up many a supper at Arrowhead. Desserts—hefty fruit puddings, pies and slumps, or cakes made with plenty of eggs, milk and butter, played a larger part in nutrition than they do today.

In the early days, cider—meaning hard cider, of course—was the universal drink on New England farms, although near the port towns, as Arrowhead was, rum ran a close second. Sometimes the two were mixed in a concoction aptly called a "Stonewall." Recent generations of Arrowhead dwellers have been teetotalers. Although this practice might date back to the Temperance Movement which began in the 1830s, we believe that it began in the 1880s or 90s. Uncle William's diary from the 1850s mentions making large quantities of cider, and we assume that the family was consuming its fair share of spirits at that time. The present younger generation has resumed the custom, although we haven't yet acquired a taste for "Stonewalls!"

As farm life has changed in the twentieth century, so have eating habits. Dick's farming day is as likely to include a hurried 100-mile round trip to pick up seed before the supplier closes at noon or a six o'clock meeting with another farmer who has a tractor to sell, as the traditional morning and afternoon work on the place. Paula fits life around a nursing schedule, and Justin's days run on school and TV cartoon times. In short, we are all more likely to eat "catch as catch can" than on the rhythmic schedule of earlier days. On Sundays, however, we often find ourselves reverting to type, eating in the older pattern and perhaps taking the time to make a favorite old dish we haven't had in a while.

The recipes which follow represent some of the favorite foods of ten generations at Arrowhead. Many of them have come down to us in faded, tattered, hand-written "receipt books" of Aunt Ida's. How far back some of the specific recipes go, we have no way of knowing, but similar versions of some of them undoubtedly graced William and Abigail's table in the 1690s. Others are from later periods, right on up to today, as each generation of cooks has adapted and modified old eating habits and introduced new ones to meet changes in tastes and the availability of foods. The tradition, like the farm, is on-going, and we've chosen to present it that way. We hope you enjoy it.

"When the oaks are gosling grey,
Plant your corn in May."
OLD FARM ADAGE

SPRING

I T   H A S been said by historians—not totally in jest—that, had other parts of the country been settled first, when we finally came to New England, we might well have considered it uninhabitable. At no time does this theory seem more plausible than in March, for spring comes late to New England. And when it does come, it is rather austere by the lights of other parts of the country, which revel in lush profusions of azaleas and spectacular displays of dogwoods. Only a New Englander can see spring in a landscape of damp brown earth barely tinged with green, in skunk cabbage and trailing arbutus, in the daily shrinkage of the last patch of snow.

Indeed, March and early April were the season of the "six weeks' want," after the contents of the root cellar had run out or gone mouldy and before the first new spring vegetables appeared. If the previous year had been a bad one, other things were likely to be wanting by then, too. Small wonder that dandelion greens and fiddlehead ferns looked good! Small wonder, too, that New Englanders have long tried to hurry spring by forcing spring bulbs into bloom in their kitchens, and bringing in boughs of forsythia, pussy willows, and other early shrubs to force as well.

When does that magic moment described on seed packets; as "when all danger of frost is past," arrive? Well, there was "the year without a summer" in 1816, when it never arrived: there was a frost every month. And in 1982, Arrowhead had 20 inches of snow on April 6. Most years, however, the land at Arrowhead has lost its snow cover by mid-March. The last frost on the upland is usually around the 20th-25th of April, and although we have had frost the 2nd or 3rd of June, most of the farm is safe from frost by the first week in May. But frost isn't the only determining factor in the spring; soil temperature is important, too. In general, seeds won't germinate until the soil temperature tells them it's safe to.

Drainage is important, too: how fast do the melting snow and frost leave the soil? Much of the land at Arrowhead is high and well drained, and we can get onto those fields to plow by late March or early April. The lowest land, on the other hand, isn't dry enough to plow until late May or early June. (This low land, however, is "strong" land, and will hold up better during the summer drought.)

The winter "slack season" ends well before plowing time. In the early days, March brought maple sugaring and fence mending as the first outdoor work. Hens had to be set in March, too, in the days before incubators; and then as now fruit trees and bushes had to be pruned. Today, March means opening up the greenhouses to start bedding plants and field transplants.

April has traditionally meant plowing and fertilizing the fields, seeding hay, and planting early crops. It also means — hallelujah! — cutting the first asparagus. The day on which the first tender stalks are brought in from the asparagus patch is undoubtedly the human equivalent of the day when the most staid old dowager among the cows frolics in the sunny green grass. Spring has arrived at last! The happy day for cows traditionally came in the old days when farmers depended on nature to provide forage, on May 20; nowadays, with seeded and fertilized pasturage, it comes in April. These days, April also brings the opening of the farm stand, with Easter plants, and, of course, asparagus. Greenhouse work, outdoor planting, and sale of spring plants continue through May and June, and Christmas trees must be planted in May.

Spring: 1890's LMP

and 1980's CMC

To the New England table, April also traditionally brought the end of the long winter of dried, salted, smoked and otherwise preserved foods. It meant the fresh taste of spring greens—dandelions, wild onions, asparagus, to be followed in May by beet greens, lettuce, and spinach; of radishes and rhubarb; of spring lamb and fresh-caught salmon; of firm orange-yolked eggs and thick, yellow cream. Lavish in quantity and quality, the vegetables and meats needed no Hollandaise or mayonnaise; their own bright taste with a bit of salt and pepper, freshly churned butter, and perhaps a touch of cider vinegar sufficed nicely, thank you. But the eggs and the cream lent themselves to dozens of subtle concoctions, from crisp golden pastry puffs filled with smooth sweet cream, and tall amber popovers drenched in melting butter, to frothy puddings topped with fluffy meringues and the never-to-be-forgotten delights of an "All-Day Cake."

FORCING BRANCHES

There comes a day in early February when winter lets down its guard. The wind has veered to the south. The air is soft and moist. We think of pussy willows, how silvery they were against the blue sky last spring, and how they melted into the atmosphere on a misty morning.

Come, it is time to steal a march on this spring. With garden pruners we visit the pussy willow tree. Yes, the buds are beginning to swell, even though the snow is swirled in a drift around the tree trunk.

We reach for branches that seem to promise the most pussies. We pick some long straight ones to add height to our bouquet, some curved ones to give it grace, and some forked ones to fill in the center, always cutting on a slant to give the bough a larger mouth with which to drink.

Then we go to the forsythia bush and clip some branches. We shall have a pitcher of silver and a pitcher of gold in our house.

Tall stoneware or pottery pitchers with slender throats and sturdy bottoms will complement our flowers and will stand firmly. Fill them with lukewarm water and plunge the branches deep because all cut boughs like a good depth of water. If you cannot arrange your branches as soon as they are picked, be sure to make a fresh cut for each stem later.

Fill your plant mist bottle with lukewarm water and mist branches now and twice a day thereafter. Indoor air seems very dry to outdoor flowers. Place branches in good light in a coolish place; 55°–60° is good. Sunshine is fine now, but not after flowers open.

How soon will they open? Wait and see. Branches respond differently in different years, depending on growing conditions of the previous summer and weather conditions they have recently endured.

Forcing flowers for a certain date is only for the experienced, and experience is gained and improved constantly by taking note how your trials respond. At best, forcing for specific dates is tricky business.

Any of the fruit tree buds may be forced if you can forego the fruit they might bear. Try apple and peach in mid-March, cherry and pear a bit later. You don't need many branches or too long stems to make very attractive fruit arrangements. We tend to be more extravagant with pussy willows and forsythia because they are abundant.

Flowering quince and lilacs also force well. Select these in early April, and look for short fat buds. Slender buds are leaves.

Children like to look on the lawn in late March or early April for tiny clumps of innocents which can be dug with a little ball of earth and grown in a small dish. On a sunny window sill they bloom surprisingly fast.

Oh, we know winter has more snow up her sleeve, but what do we care? We have the promise of spring, and we shall turn our backs and watch our buds open.

SPRING GREENS

DANDELION GREENS

Little folks always looked forward to the spring dandelion expedition with Auntie as one of the first rites of the season. A heady scent arose from the earth as one stooped with the knife to gather these treasures. Auntie knew all the south-facing nooks where the bravest plants emerged first, and an hour's walk about the place usually yielded a fine basketful.

In the kitchen the greens were carefully rinsed, tossed into a covered kettle, and cooked with very little water for 15-25 minutes, depending on age and tenderness. They came to the table as a dark green mound with butter melting over them.

ASPARAGUS

The sun is shining, the landscape newly green and the late April breeze is mild. Surely this is the day! Equipped with basket and sharp knife we sally forth—across the mowing lot, through the apple orchard where the kingbirds live, down to the asparagus field. From a distance it still looks bare and brown. But we are not deceived. Perseverance pays, and we hurry from plant to plant, cutting one here, none there, and then two more. Finally, we have enough for everyone.

In the kitchen, after removing the hard butts and rinsing, we put the tender stalks and tips into boiling water and cook 10-12 minutes, meanwhile preparing some buttery toast.

Some like the tips on toast, and some like a poached egg on the toast with a goodly bunch of asparagus beside it. Some want only asparagus. Glen loved even the water it was cooked in, sipping it from a cup and calling it his tea. There are many things to do with this green vegetable but Arrowhead loves it best plain or on toast.

SPINACH

In the past spinach has been much maligned and greatly overcooked. The new varieties grow more quickly and are ready in early spring. Pull the young plants shortly before using, rinse carefully to eliminate any sand, place in saucepan with only a bit more water than clings to them and simmer over low heat for about 5 minutes. Your spinach will come to table

showing the form of its leaf, thus appealing
to the eye, and the tender sweet mouthful
will appeal to the appetite.

Some people cook only the leaves, which
produces a distinctively delicate flavor, but
we feel there is more value and succulence
to using the whole plant, stem and all. Enjoy with butter or vinegar or both.

BEET GREENS

Beet greens grow in sprightly little rows in
the "table garden," and if the row is properly thinned out as you pull your greens, the
remaining plants turn into full-fledged beets.
These first baby greens come to table with no beets attached. Tender and
succulent, they require only 10 minutes cooking. Use little more water than
clings to them after rinsing, and boil slowly to release their sweet goodness.

Soon the greens will appear with larger leaves, longer stems and tiny
round beets. These will need about 20 minutes cooking.

After this stage you will boil the beets and their tops separately, vary-
ing cooking time according to size of beet and texture and age of leaf.

Enjoy all stages hot or cold with butter or vinegar.

PAULA'S CREAM OF ANYTHING SOUP

2 tbsp. flour   *2 tbsp. butter*
10 oz. fresh whatever   *1 qt. chicken broth*
1 small onion, chopped   *(homemade, of course)*
salt and pepper   *½ cup heavy cream*

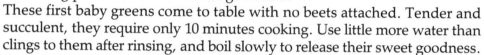

Melt 2 tbsp. butter, add flour, and cook 3 minutes over low heat. Chop
your whatever (broccoli, asparagus, spinach, cauliflower). Saute onion in
butter. Stir in vegetable, cook 3 minutes. Stir in broth and heat to simmer
(now you can make the chicken salad out of the poached chicken). Add
cooked roux to soup, stirring until smooth. Simmer until vegetable is
tender. Cool. Puree in blender and return to pan. Add cream and season
to taste. Add some reserved vegetable if desired. This soup is best if made
a day ahead.

RHUBARB

Rhubarb has been popular in New England ever since Benjamin Franklin sent back the first seed from England in 1760. Just when everybody was tired of apple pies and sauce—and the apples had about run out—along came rhubarb, tart and fresh and juicy. Where the custom of combining it with strawberries came from we're not sure, but it certainly wasn't Arrowhead! Rhubarb is at its best in May, and people on the farm generally stopped cutting it at the end of the month, well before strawberries were ripe; so never the twain met.

RHUBARB PIE

6–7 medium stalks rhubarb   *2 cups sugar*

Make pastry for 9" pie plate, 2 crusts. Chop rhubarb rather fine, removing any tough outer strings. Place in lined pie plate—you should have about 3 cups rhubarb. Sprinkle a little flour over it. Then cover it with the sugar. Again sprinkle with a bit of flour, then add about 1/2 cup water. Add top crust, slashing it in several places.

Bake in preheated oven 450° for 10 minutes, then reduce heat to 375° and bake about 40 minutes more or until pie is bubbling in center slit and crust is golden.

PIE CRUST

Pastry is a very personalized creation, perhaps even reflecting the character of its creator. Two cooks using the same recipe usually get slightly different results—sometimes quite different. It is generally recognized however, that the approach to any pastry should be light with the hand, the blending implement, and the rolling pin. Thinly rolled pastry usually makes superior pie crust. In rolling out the dough, never roll the pin in exactly the same direction twice in succession and your dough will be less likely to split. When ready to slide rolled pastry from board to pie plate fold it in center and pull the half circle thus formed halfway across plate. Then unfold, fit, and trim.

Here three generations of Arrowhead cooks offer you their best pastries, all made many times on the same board: All three cooks stress the importance of *icy* cold water, and all three also prefer an unbleached high-gluten flour such as King Arthur's.

AUNT IDA'S PASTRY

This recipe makes two 2-crust pies and one pie shell; it comes from the days when no one *ever* made just one pie.

1 tsp. baking powder	1 cup butter
pinch salt	1 cup cold water
5 cups flour	1 cup lard

Sift together baking powder, flour, and salt. Work in shortening lightly with fingertips until crumbly. Add cold water gradually, cutting in after each addition. Gather dough together and divide into five parts. Place portion on floured pie board and sprinkle top with flour. Roll out to about 1/8" thickness. Line pie plate and moisten rim of pastry with cold water. Put in filling. Roll and fit top crust, pressing edges lightly with thumbs to seal in filling. Pick top with fork in several places and slash vent in center. Bake in preheated oven 450° for 10 minutes, then lower heat to 375° for 40–45 minutes more.

A single crust requires slightly lower temperature, so bake 10 minutes 400° and lower to 375° for about 15 minutes.

CHARLOTTE'S PASTRY

This recipe was evolved to fit into a low-sodium, low cholesterol diet. It makes a tender crust with a distinctive slightly sweet flavor all its own, which was so well received it has now become a standard favorite.

Be sure to use the specified margarine, as others do not perform as well.

2 cups flour	½ tsp. baking powder (or
½ cup Fleischman's soft	1 tsp. featherweight
margarine	low-sodium baking powder)
	Scant ½ cup ice water

Mix together flour and baking powder. Add soft margarine and rub lightly together with fingers until coarsely crumbled. Cut in water, one tbsp. at a time, with sharp knife. Use fork to gather dough into 2 portions and proceed as above.

POLLY'S PASTRY

This makes a light, crisp pastry.

2 cups flour   ⅔ cup solid vegetable shortening
pinch salt   ice water as needed

Sift flour and salt together. Cut in shortening, using two table knives held together. Add just enough ice water to hold dough together, mixing lightly with fingers. Form into two portions; proceed as above.

In the old days, farm families kept hens around for a second year, even though they would not lay as heavily, in order to have a supply of chicken for the table. When one was wanted, father went out to the hen house and selected one whose pale-colored comb indicated she was through laying for a while. This supplied the table with chicken while reducing the egg supply as little as possible. Of course, "chicken" was something of a euphemism in this case, and fricasee was a good use for the older, tougher hens.

CHICKEN FRICASEE

Buy a whole chicken and have it cut into pieces, as freshly cut-up chicken is more flavorful. Place parts in deep cooker and cover with water. Add one stalk celery with leaves and one medium onion quartered. Cover cooker and simmer about 1½–2 hours or until meat comes away from bones easily. Remove skin and bones and set meat aside.

You should have about 1 quart of chicken stock. If more, reduce by boiling; if less, add water. Work 2 tbsp. flour into a smooth paste with 2 tbsp. cold water. Blend this into chicken stock; add 1 tbsp. butter. Stir while stock thickens a bit and simmer 15 minutes longer. Salt and pepper to taste. Add chicken pieces and hold on warm for flavours to blend.

Serve hot on buttered baking powder biscuit halves. Make these with shortcake recipe (Charlott's, page 114). Roll out on floured board to about 1/2" thickness and cut into rounds. Bake as for whole shortcake patty.

This is an especially good dish on a chilly spring evening.

EASTER MENU

Baked ham/cloves
Riced potatoes
Spinach*
Parsnips
Parker House rolls*
Neapolitan Pudding*
Snow Pudding*
Coffee

The traditional Easter menu at Arrowhead made use of the seasonally available vegetables and the traditional ham. Like other meats, ham was cooked very plainly, with no sauces or glazes; just cloves and perhaps a bit of brown sugar. Potatoes, which were usually somewhat heavy by this time of year, were often riced to lighten them, then served with plenty of fresh, sweet butter. Fluffy Parker House rolls were part of many festive occasions, and always on the table at Easter. For desserts, there was a light, creamy pudding, perhaps a Snow Pudding or Neapolitan Pudding.

Aunt Lizzie's Easter Buns were served as a dessert at Easter Supper.

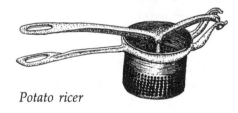

Potato ricer

AUNT LIZZIE'S EASTER BUNS

1 yeast cake	2 cups milk
¼ cup water lukewarm	1 tsp. salt
2 cups sugar	1 tsp. nutmeg, freshly grated
½ cup butter	7½ cups flour
1 egg	1 cup currants

Dissolve yeast cake in lukewarm water.

Scald milk and set aside too cool; when lukewarm add the dissolved yeast. Meanwhile cream butter and sugar together and add the beaten

*See our recipes.

egg. Sift together flour, salt, and nutmeg. Combine milk with butter and sugar and stir in flour mixture gradually, mixing well after each addition.

Turn onto floured board to knead. If necessary add more flour for proper consistency to handle. Knead thoroughly, let rise overnight. Knead again, mould into flat cake and fold in three pieces. Knead and fold in this way four separate times. Add the lightly floured currents. Cut off small pieces and shape into rolls. Brush with butter, place close together in pans and let rise. Bake in oven 375° thirty minutes. Brush with 3 tbsp. sugar dissolved in a little water. Return to oven until glaze is dry.

PARKER HOUSE ROLLS

2 cups scalded milk	¼ cup butter melted
½ yeast cake	2 tbsp. sugar
¼ cup lukewarm water	1 tsp. salt
	2 cups flour

Mix scalded milk, when cool, with yeast cake which has been dissolved in lukewarm water. Add flour; beat thoroughly and let rise.

When spongy add remaining ingredients and enough flour to knead. Knead and let rise to double its bulk. Then shape into balls, lay on buttered sheet and cover with large pan. When risen to double their bulk, press with handle of wooden spoon, almost dividing biscuit. Brush one half with butter; press the two halves together. Place on buttered tin, let rise. When light, bake for 10–15 minutes. Moderate oven, 350°.

We can't find Aunt Ida's recipe for this favorite spring dessert, but the 1897 *Cookbook of the Women's Auxiliary of the Amesbury YMCA* gives the following instructions for "Neapolitan Blanc Mange":

NEAPOLITAN PUDDING

2 envlopes gelatin	1 tsp. vanilla
1 qt. milk	1 egg yolk, beaten
¾ cup sugar	1 sq. chocolate, grated

Soak the gelatin 5 minutes in 1/2 cup of the milk. Scald remaining milk, add sugar and softened gelatin and stir to dissolve. Remove from heat and divide into three parts. Flavor one with the vanilla; color second part with beaten egg yolk; and color third part by stirring in grated chocolate. Set away until cool and slightly stiffened. Pour white part into mold; chill. Add yellow part; chill. Add chocolate part; chill thoroughly. Serve with whipped cream.

SNOW PUDDING

1 envelope plain gelatine   *juice of 1 lemon*
¼ cup cold water   *¾ cup sugar*
1 cup hot water   *1 egg white*

Soak gelatine in cold water 5 minutes. Add boiling water and sugar, stir until dissolved, add strained lemon juice and chill until it begins to harden. Now add stiffly beaten egg white, beating whole mixture together until it becomes spongy and has increased in volume. Chill until firm. Serve with custard sauce.

CUSTARD SAUCE

1 egg yolk   *½ tsp. cornstarch*
¼ cup sugar   *1 cup milk*

Beat together egg yolk, sugar, and cornstarch. Scald milk and combine with egg yolk mixture in double boiler. Cook just until it thickens, stirring constantly. Flavor with vanilla and let cool. Serve over pudding.

We also occasionally make this pudding, when our potted tree cooperates, with sour orange. Different and delicious!

Like iced coffee and coffee ice cream, coffee gelatine is more appreciated in New England than in other parts of the country. It makes a lovely cool ending to a meal.

COFFEE GELATIN

1 envelope plain gelatin   *1½ cups hot coffee*
½ cup cold coffee   *½ cup sugar*

Coffee should be fairly strong. Soak gelatin in cold coffee 5 minutes. Add hot coffee and sugar and stir till dissolved. Chill until firm and serve with whipped cream or pouring cream.

LAMB

Lamb, a meat not properly appreciated in some parts of the country, has always been a staple in New England, and we're sure it has been served at Arrowhead since the first William began to raise sheep on his four acres. While lamb is also tasty on the pink side, the New England way is well done, and if properly cooked, it will remain juicy while acquiring a rich, browned taste. Mint jelly, we might add, is a modern innovation. Prepared in the following way, lamb is just fine all by itself.

ROAST LEG OF LAMB WITH POTATOES AND LAMB GRAVY

1 leg spring lamb, about 8 lbs.   *6–7 medium potatoes, pared and*
1 quart water   *cut into 5–6 pieces each*

Place lamb leg in uncovered roaster pan with 1 qt. water and put into preheated 450° oven. Cook 1/2 hour then lower heat to 375° and put cover on roaster leaving vent open to gradually evaporate water. Roast 1 1/2–2 hours, remove cover, and let juices dry and brown on pan while you prepare potatoes.

When the pan is nicely browned add 6 cups water and stir in thoroughly to make brown juice. Place potato chunks around lamb and continue slowly roasting for 3/4–1 hour longer, occasionally basting with brown juice.

When meat shrinks from bone remove it and potatoes to keep warm while you make lamb gravy.

Stir in flour to the proportion of 1 tbsp. to each cup juice. If need be add a little more water. You should end up with about 3 cups or a quart of really super gravy.

Ironstone gravy tureen

LAMB STEAKS

Lamb steaks are a delicious variation from lamb chops and roast lamb.
Have the butcher cut 3/4–1″ slices from the face of a leg of lamb. Pan fry over medium high heat until well browned on both sides.

GROUND LAMB ON TOAST

Grind some of the cold lamb left from your Sunday roast with medium blade of meat grinder. Put about 2 cups into a saucepan with 1/2 cup lamb gravy. Heat slowly, stirring with fork, and bring to desired consistency by adding a little water. The addition of water gives meat a lighter texture. Spread on hot buttered toast slices.

Served with either one hot vegetable or a side dish of cold peach preserves, it makes a quick and delicious lunch.

In the days before refrigeration, and when people were dependent on their own flock of hens for all their eggs, this recipe assured a steadier supply. This was often done in late spring, to prepare for the summer molting season, when production dropped.

TO PUT DOWN EGGS IN LIME WATER

6 qts. water   ½ pint Airslacked Lime
                          ¼ pint salt

Put in crock or jar and stir well. Let stand overnight. Stir well. Be sure crock is half filled. Put in eggs, being careful that all are fully submerged in lime water. Preserve when eggs are plentiful and low in price, and use for cooking when eggs are scarce.

March brought a renewed supply of eggs to the farm kitchen after the cold winter when the hens nearly stopped laying. Thus many of the recipes we associate with spring use an abundance of eggs, and they seem to go together especially well with spring greens, as well as with the Easter ham. They are used in many ways: alone, as thickening for a variety of custard puddings and sauces, or as a rising agent in breads and cakes.

SOFT BOILED EGGS

2 eggs boiled 3½ minutes

The magic formula for a super soft-boiled breakfast:
   Lower eggs gently into boiling water, lest you crack the shell and the white escape. Then count the minutes and seconds. Now, a smart crack with the knife, scoop out the innards into a blue bowl (wherein the taste is greatly enhanced!), a dab of butter, a dash of pepper and salt, and presto—serve it to the king.

OLD-FASHIONED PUFFY OMELET

4 eggs	1 tsp. baking powder
½ tsp. salt	1 tsp. cornstarch
dash of pepper	½ cup milk

Beat whites and yolks separately. Stir cornstarch, salt, pepper, and baking powder into milk and mix with beaten egg yolks. Add the egg whites beaten light and fluffy, and fold in quickly and thoroughly. Pour into hot buttered frypan and cook slowly until well puffed up and lightly browned beneath. Put in 350° oven to dry slightly and form thin skin on top. Fold in half with spatula and serve immediately from a hot platter.

Notes on French omelet-making: A special pan reserved only for making omelets makes the process *much* easier. Season a new pan according to instruction. After each use, wipe inside with a dry paper towel instead of washing. These omelets are the exception to the general rule about cooking eggs slowly. High heat makes a light, fluffy omelet with a golden, crispy crust. If the butter should brown slightly, it won't hurt. If it burns darkly, discard it, wipe the pan clean, and start again.
   Never use more than 3 eggs in a French-style omelet, as it changes the texture. It takes only a couple of minutes to make each one, and the results are worth it.

POLLY'S SPINACH AND CHEESE OMELET,
FRENCH STYLE

For each omelet you will need:

2 eggs   *1–1½ tbsp. generous shake of*
3–4 oz. fresh spinach   *Parmesan cheese*
3–4 tbsp. cottage cheese   *salt*
                          fresh ground pepper

Wash spinach, shake off excess water. Place in saucepan over medium heat with only the moisture that clings to the leaves and allow to wilt slightly. Stir in cottage and Parmesan cheese and a dash of salt, and heat through. Do not allow cottage cheese to liquefy—remove from heat if necessary. Meanwhile, preheat omelet pan over high heat, and beat eggs with salt and generous grinding of pepper. Melt 1 tsp. butter in hot pan. When it stops foaming, pour in eggs, swirling pan to coat bottom. Cook over high heat, shaking pan gently to distribute egg evenly, to desired firmness.

(Bottom should be a golden brown and center slightly runny.) Slide omelet onto warmed plate, heap spinach mixture along center and fold edges over. Serve immediately.

CUP CUSTARDS

3 eggs, beaten well   *2 cups milk*
½ cup sugar   *1 tsp. vanilla (optional)*

Beat all together and fill your best custard cups. Sprinkle fresh grated nutmeg on top. Bake in shallow pan of water with oven at 350° for about 45 minutes, or until inserted knife comes out clean.

CHILDREN'S EGG NOG

1 egg   *2 tsp. cocoa*
2 tbsp. sugar   *1 cup milk*

Separate egg, beat white until foamy, add sugar and beat until stiff. Dissolve cocoa in bit of hot water and beat into milk together with egg yolk. Pour over egg white and stir in with spoon. Serve in tall frosty glass. Watch it disappear!

So many pancake recipes make enough for an army, we thought you might enjoy our formula for breakfast-for-two. It can, of course, be multiplied for more.

Spring, when the new maple syrup was at its delicate best, has always called for plenty of pancake breakfasts—and sometimes for pancake suppers.

PANCAKES

¾ cup flour	1 egg
1½ tsp. baking powder	¾ cup milk
1 tbsp. butter, melted	

Sift dry ingredients together; add beaten egg, milk, and shortening, stirring all together well. Cook in lightly greased, moderately hot pan. Turn only once.

Pancake batter should be thin, the cakes should be dropped in small spoonfulls to make dainty thin pancakes. The pan should be not too hot, so the cakes cook rather slowly, thus insuring a golden brown, slightly crusty finish. Butter while hot and enjoy with maple syrup and a steaming cup of coffee.

The following popover recipe is from the original 1896 Fanny Farmer. We substitute two minutes at low speed of electric mixer for the Dover egg beater, but otherwise follow it religiously. Sarah always achieved mountainous popovers in a tin muffin pan, but we find a cast iron one is really best, and worth the investment if you're serious about popovers. If you don't have an iron one, place the "hissing hot" aluminum one over low heat while you fill the cups, to maintain the temperature. Fill cups two-thirds full, and pop into 450° oven immediately. Turn back to 425° and *don't* peek for at least 30 minutes. Popovers are tricky, and you may not get puffy ones the first time or two. But once you get the knack, they are wonderfully simple—and simply wonderful!

POPOVERS

1 cup flour	⅞ cup milk
¼ teaspoon salt	2 eggs
½ tsp. melted butter	

Mix salt and flour; add milk gradually, in order to obtain a smooth batter. Add egg, beaten until light, and butter; beat two minutes, using Dover eggbeater. Turn into hissing hot buttered iron gem pans, and bake thirty to thirty-five minutes in a hot oven. They may be baked in buttered earthen cups, when the bottom will have a glazed appearance. Small round iron gem pans are best for popovers.

Charlie especially liked these when he and Sarah were courting.

SARAH'S CREAM PUFFS

½ cup butter melted in
1 cup boiling water in small saucepan.
While boiling stir in
1 cup flour. Remove from heat.
When entirely cool, add
1 tsp. baking powder and
3 eggs added one after the other
without beating beforehand.

Drop by spoonfuls on buttered tin and bake in hot oven 450° for 20 minutes. Let them stand on the tin to cool.

CREAM FILLING

1 cup milk	*1 tsp. butter*
½ cup sugar	*1 egg*
2 tbsp. cornstarch	*1 tsp. vanilla*

Smooth cornstarch into sugar with spoon. Stir into milk in double boiler and continue stirring while it thickens. Add butter and cook 5 minutes. Then stir in beaten egg and cook 2 minutes longer. Add vanilla and cool before serving in cakes.

Tapioca comes from the root of the bitter cassava, native to South America, and came into New England with the Caribbean trade. The larger "pearl" form is the old-fashioned form, and like many old-fashioned things, has a more distinct character than the later "minute" tapioca. Practically off the market at one time, this tapioca is now found in natural food stores and "country stores."

Originally named Baked Pearl Tapioca, this pudding was re-christened by Glen who was very fond of it—even though he considered most puddings "ladies' food."

"FISH EYE" PUDDING

⅓ cup large pearl tapioca	*2 eggs*
¾ cup water	*½ cup sugar*
2¼ cups milk	*½ tsp. vanilla*
½ tsp. salt	

Soak tapioca in the water for 1 hour. Then discard remaining water and add milk and salt. Beat eggs and sugar together; stir into mixture with the vanilla. Pour into ironstone or pyrex baking dish and put into 350° oven, placing dish in pan with a little water. Bake about 2 hours, stirring a couple of times the first hour. Pudding is done when an inserted knife comes out clean. Excellent served cold as well as hot. No sauce needed.

We find that different brands of pearl tapioca respond differently to soaking, some requiring less time than others. The finished pudding is best when the pearls are separated and a little firm to the tongue. So experiment a bit to find how your pearls respond to water.

A treat any day, Floating Island was always made as a special appetizer for those recovering from a bout with the flu or other insidious malaise.

FLOATING ISLAND PUDDING

2 small or 1 large
   California orange
2 tbsp. sugar
2½ cups milk
3 tbsp. cornstarch
½ cup sugar

1 tsp. butter
1 whole egg and
   2 egg yolks
bit of salt
½ tsp. vanilla
2 egg whites

4 tbsp. sugar

Cut orange in half, cut around sections, scoop fruit into bowl with juice, add 2 tbsp. sugar, and set aside.

   Scald milk in double boiler. Dissolve cornstarch with bit of milk and add to scalded milk with sugar and butter. Stir while it thickens, then add beaten egg and egg yolks and cook not more than 3 minutes, stirring constantly. Flavor with vanilla. Pour through strainer into pyrex or pottery baking dish. Let cool. Fold in orange, mixing very little.

   Beat egg whites until very stiff, then beat in sugar. Drop by spoonful on top of custard mixture. Place in 375° oven until meringue islands are sufficiently baked.

NEVER FAIL SPONGE CAKE

2 eggs beaten 5 minutes
Add *1 cup sugar* and beat 2 minutes
Stir in *1 cup flour* sifted with *1 tsp. baking powder* and
1 tsp. butter
Flavor with 1 tsp. vanilla

Pour into round greased cake tin 8″ and shallow. Bake 375° oven about 20 minutes, and remove while it is still "singing" slightly. This is a good basic sponge cake of many uses.

WASHINGTON PIE

Split the cake when cool and spread raspberry jam on bottom layer and pile whipped cream on top just before serving.

DAFFODIL CAKE

WHITE PART

6 *egg whites*, beaten stiff
½ *tsp. cream of tartar*, added to whites when foamy
pinch salt
½ *cup flour* and ¾ *cup sugar*, sifted together four times
½ *tsp. vanilla*

Add flour and sugar mixture to beaten egg whites gradually, folding it in with large spoon after each addition. Set aside.

YELLOW PART

6 *egg yolks*, beaten 3 minutes
¾ *cup sugar*, add and keep beating
¾ *cup flour*, sifted 4 times with 1 tsp. baking powder
¼ *cup boiling water*, added last
½ *tsp. vanilla*

Add flour and baking powder gradually to egg yolk mixture, stir in after each addition. Add boiling water all at once and quickly beat it in.

Pour white part into large ungreased angel cake tin. Then pour in yellow part on top without mixing. Bake in 350° oven about 40 minutes, or until it is through "singing." Cool and frost with confectioner's sugar frosting.

This is the official birthday cake of Arrowhead.

DELICIOUS LAYER CAKE

Also known as "All Day Cake." It *does* take time, but is well worth it! Aunt Ida occasionally baked this cake on special order, and her customers happily paid $3.50 for it in the 1920's. The recipe probably dates from the late nineteenth century, when "ribbon" cakes of vari-colored layers were popular.

WHITE

1 cup sugar
¼ cup butter
½ cup milk
whites of 4 eggs

1½ cup flour
½ tsp. soda and 1 tsp. cream of tartar
(or 1½ tsp. baking powder)
¼ tsp. vanilla

Cream butter and sugar. Add milk alternately with flour, which has been sifted with soda and cream of tartar. Fold in stiffly beaten egg whites and vanilla.

YELLOW

1 cup sugar
2 tbsp. butter
½ cup milk
yolks of 4 eggs
1½ cups flour

½ tsp. baking soda
and 1 tsp. cream of tartar
(or 1½ tsp. baking powder)
¼ tsp. vanilla

Cream butter and sugar; beat in egg yolks one at a time. Add vanilla. Add milk alternately with flour, which has been sifted with soda and cream of tartar.

CHOCOLATE

Over low heat, melt 1 square baking chocolate with 2 tbsp. sugar. Combine 1/3 of white batter and 1/3 of yellow batter in mixing bowl; add chocolate and mix thoroughly.

Pour each batter into greased and lightly floured 7" x 11" rectangular or 9" square pan. Bake 350° for 20 min., or until center springs back when touched. Turn out onto racks and cool.

FILLING

Dissolve 1 square baking chocolate in 1 cup boiling water; add 3/4 cup sugar and 1 tbsp. butter. Let come to a boil, then stir in 1 heaping tablespoon cornstarch which has been dissolved in 1/2 cup cold water. Reduce heat and cook, stirring, until thick and smooth. Add 1 tsp. vanilla and remove from heat. Spread over yellow and chocolate layers while still hot. Allow filling to cool and set slightly before stacking chocolate on yellow, white on chocolate layers.

FROSTING

Cream 1 tbsp. butter with 1/2 lb. confectioner's sugar. Add milk to spreading consistency. Frost top only.

Lemons were a rare and special treat in early New England. They came from the Mediterranean countries where Yankee ships sold salt fish, and were one of the luxuries which were more readily available near the port towns.

LEMON MERINGUE PIE

1 cup sugar	1 cup boiling water
1 tbsp. flour, rounded	1 tsp. butter
juice of 1 large or	1 whole egg and
2 small lemons	yolks of 2 more

Cover 8" pie plate with pastry. Prick all over and bake creamy golden. Blend flour into sugar and add boiling water, butter, beaten egg and yolks, and lemon juice. Stir in double boiler while it thickens. Cook about 10 minutes. Pour into crust and let cool a bit. Frost with meringue.

    2 limes instead of the lemons turn this into a delicious lime pie. Or, if you happen to have an ornamental orange tree, 2 or 3 make a delicate, and different pie.

MERINGUE

2 egg whites	4 tbsp. sugar

Partially whip whites, then gradually add sugar. Beat until stiff. Bake in 350° oven to a nice golden color. Cool slowly. Refrigerate.

"As the days grow shorter,
The weather gets hotter."
OLD FARM ADAGE

SUMMER

"TO BE New England is to get up early, and make the most of the day," goes an old saying. Being "New England," folks at Arrowhead have always believed wholeheartedly in this dictum, having little patience with slug-a-beds. But in no season is this precept more apt than in summer, although today, as always, it has different meanings to different people at Arrowhead, depending on their responsibilities. To some of us, it means falling out of bed at three, making the 40-mile run, fueled by all-night-diner coffee on I-95, to the already-bustling Chelsea produce market, transacting a round of business and hurrying back in time to get the day's operation started on the farm. To others, it means arising when the air is clear and cool and the world is quiet, and beginning the day with a simple breakfast on the piazza. The mourning doves call gently in the pines, the fuchsias and begonias in the porch pots stand fresh and bright in the early morning sun, and a cool breeze sweeps across the field. Then

Summer: 1890's. LMP

Summer: 1970's. CMC

the world can be enjoyed, and the day planned, before the heat and commotion commence, for there is much work to be done.

From the bright, jewel-like days of June through the blazing heat of July and the sultry dog days of August, work continues at a fever pitch on a New England farm. Planting and tillage continue; harvesting begins in earnest. It is a time of much work, but also many rewards.

More often than not, summers at Arrowhead mean drought. Much of the land is high and gravelly, and while that allows early plowing and early crops, it also means that crops burn up fast in a dry spell later on. This droughty land, which was best used for forage, made livestock traditionally important at Arrowhead. Today, for raising table crops, it means careful planning to plant early crops and drought-resistant summer ones on the high ground, and later, thirstier crops in the stronger low lands.

Conversely, rain and dampness can be a problem in the summer, too. Thundershowers tend to follow the river, and many a haying crew has had to race to get the last load in before the thunder headers rolling in from the Northwest reached Arrowhead. And if they failed, the hay took longer to dry out again because of the damp sea breezes from the nearby ocean. These breezes also keep the temperature down, making summer crops slightly later than in inland areas not far away. These same breezes, however, extend the growing season in the spring and fall, when they are a warming influence on the cold land. And they are most welcome on a sweltering July day, when a refreshing gust from the east, carrying a faint whiff of salt, announces that there has been a "sea turn," and the worst of the heat wave is over.

From the days when the first William and his sons swung their iron-bladed scythes to gather the natural meadow and marsh grass to today, when the big green John Deere tractor pulls a heavy load of rectangular bales toward the barn, June has brought haying time, with its hot, sticky, prickly work, and its dry sweet scent. No longer, however, is haying the home place fields followed by haying the marshes. In William's time, and up into Charlie's youth, July meant strapping bog shoes on the horses to cut the salt marshes, cocking the hay on staddles; and finally, "going the freight" downriver with the gundalow to bring it home. Today we buy salt hay from farmers in Rowley and Newbury, who still hay the marshes, to use for winter cover on strawberries and other perennials.

June brings other tasks besides haying: cultivating to keep down the weeds and to aerate the soil, planting later crops, and harvesting others — strawberries, peas, and, at the end of the month, the first zucchini. The field work of tending and picking continues in July with sweet corn, raspberries, peas, beets, beans, new potatoes; the list is endless. In August, the drying of flowers begins.

Meanwhile, in the kitchen, the arrival of all these crops brings a flurry of preserving activities. In the old days, it meant drying or canning; today it often means freezing in addition to the traditional pickling and jam- and jelly-making. Whatever the method, the cleaning, washing, husking, shelling, peeling, and slicing proceed apace, and the good smells of summer drift out of the steamy kitchen.

But summer is not all work at Arrowhead. When time can be stolen, it also means outdoor dinners and suppers, ranging from the simple expedient of plates carried out to the Gloucester hammock or to lawn chairs, to full-fledged cookouts and tables set on the piazza, and picnics. At home under the pines in the back pasture or along the bank of the river, or away — at the beach, on the rocks at Cape Ann, or along the coast of Maine — picnics have long been a favorite summer pastime at Arrowhead. For this reason, easily portable foods are summer standbys.

The primary foods of summer, of course, are the profusion of fresh fruits and vegetables. Some, like strawberries, raspberries, and peas, have relatively fleeting seasons, making them even more special. Others, like salad greens, tomatoes, beets, summer squashes, and beans, are there all summer, to be savored at will. Partly because their own flavors are so delightful, and partly because the cook likes to spend as little as possible of these lovely summer days in the kitchen, meals built around these fruits and vegetables are often simple. With this garden bounty, cold sliced meats taste especially good. They can be cooked early in the day and then served at several meals with no further hot kitchen chores.

The fruits and plentiful cream of summer lend themselves to a tempting array of cold, light, and tasty desserts. We have often wondered whether the existence of summer boarders at Arrowhead for many years delights, or whether the prevalence of such concoctions as "Charlotte Rouche" and "Summer Washington Pie" attracted the summer boarders. Whichever was the case, we are happy to be the benefactors of a lovely legacy of summer desserts.

LETTUCE

Those who think California Iceberg is the only lettuce are missing a delightful treat!

Lettuce does not like the heat of midsummer, but in late spring and again in the fall, we grow and offer our customers nine varieties to choose from. Salad Bowl, Buttercrunch, Ruby, Dark Green Boston, Grand Rapids, Waldman's Green, Romaine, Montello, and Mesa—each variety has its own devotees.

Some varieties like Salad Bowl, Ruby, and Romaine give different taste and texture to salads, and stand up well to zesty dressings.

Boston and Grand Rapids are two varieties that have been grown at Arrowhead for many, many years, and these we still often eat in the old country way, with vinegar and sugar.

HONEY

"A swarm of bees in May
Is worth a load of Hay.
A swarm of bees in June
Is worth a silver spoon.
A swarm of bees in July
Isn't worth a fly."
OLD FARM ADAGE

This old poem reflects the fact that early farmers captured swarms of native bees to produce honey. A swarm captured in May had time to produce plenty of honey for the farmer as well as enough to keep themselves alive through the winter. A swarm captured in June had less time to make surplus for the farmer, and a swarm captured in July might not even have time to produce enough to keep themselves alive through the winter. But times have changed for bees as well as for people.

Today, our farm honey bees are really a well-travelled lot. In May they

Cape Cod, and then travel to Washington County, Maine to pollinate the blueberry crop. For the summer and fall they are residents of Arrowhead, and we bottle and market our own honey from about 300 hives.

In late fall, traveling by tractor trailer, they head south to the sunshine state, to spend the winter working the citrus groves, an arrangement that makes the farmer envious!

MOLASSES GINGER WATER

2 qts. water	½ cup molasses
½ cup sugar	2 tbsp. sharp vinegar
1½–2 tsp. ginger (or to your taste)	

Stir all together thoroughly.

This is an old, old haying time beverage. To the hand who has been pitching up, or the hand who has been taking away the hay and laying the mow in the top of the barn on a sultry July day, this drink was heartening. It was always made with cold water from the well or kitchen faucet and no ice allowed, as haying is such a heating business that neither man nor beast should drink extremely cold water or too much water until he has cooled off a bit. The ginger and vinegar were in themselves cooling, and the sugar and molasses renewed energy for the next load. Here's to the ginger water!

"Up country"—in New Hampshire and Vermont—this drink is known as switchel, but "down country" in Massachusetts we've never heard it called anything but molasses ginger water or just ginger water.

In summer, when the cows gave plenty of milk, there was likely to be some milk and cream that went sour—especially in the early days before ice chests. Imbued with the "Eat it up . . ." precept, New England cooks developed many delicious ways to use this surplus. Today's pasteurized milk and cream do not sour properly, but develop a bitter taste. So for recipes calling for sour cream, we use commercially soured cream. For those calling for sour milk, we substitute either a combination of 2/3 sour cream and 1/3 skim milk, or regular milk "soured" by adding 1 tsp. of vinegar per cup, and allowing to stand at room temperature for several hours. None of these works properly for cheese-making, however.

GREAT-GRANDMOTHER'S CREAM CAKE

Break 2 eggs in a measuring cup; fill with thick sour cream. Mix into 1 heaping cup of sugar and beat until smooth. Then stir in 1 1/3 cups flour sifted with 1/4 tsp. soda. Bake in cupcake cups, oven 350° for 20 minutes.

This is an old, old recipe and just as delicious today as in the 1700s.

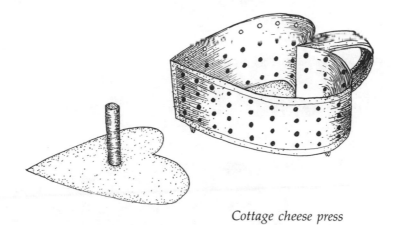

Cottage cheese press

On a certain summer day bright eyes at Aunt Lizzie's elbow would see her take the little heart-shaped cheese press from the pantry shelf, and know that a fine supper of cottage cheese was in the offing; perhaps there would be garden lettuce and hot biscuits with sliced strawberries to go with it.

For those of you who have a pet cow, and are therefore blessed with raw milk which sours properly, we print this ancient process as follows:

COTTAGE CHEESE

Set 2 quarts freshly soured milk in a pan placed on a trivet with extremely low heat under it. Pan must not have direct contact with heating unit. Let milk heat very slowly to 100°. If you lack a thermometer, let your finger tell you when the bonny clabber is just a crumb more than lukewarm. Now place your cheese press in a large milk pan to catch the whey as it drains, place the clabber therein and let set 2 or 3 hours.

Season the curd that is left with softened butter and a little salt and beat it well until curd is thoroughly broken up and seasoning distributed. Form it into a mound and place in the covered pressed-glass cheese dish, ready to chill and serve.

Elizabeth off to pick berries c.1913 LMP, *and, below, Cousin Elsie, c.1897* LMP

STRAWBERRIES

Arrowhead has good land for strawberries, and they have long been an important crop here. At the end of the nineteenth and beginning of the twentieth centuries, Charlie's strawberries were well known in Newbury-port for their high quality.

Today, strawberries are still sort of a special crop at Arrowhead. Having once been told by his father that strawberry culture was likely beyond his abilities, the grower probably tries a little harder with this crop than with some others.

And strawberries really are an effort. Planting is done in early April, and 15 months will pass before the harvest—months of weeding, cultivating, spraying, winter mulching with salt hay, spring mulch removal, irrigating, frost protection, and an occasional silent prayer or not-so-silent expletive! But every June provides a great return on these efforts, as people turn out en masse to pick the tasty red berries.

They're a finicky and costly crop to grow, but the per acre return is high, and a good crop is a great satisfaction. We have several favorite varieties, among them Raritan, Guardian, Fletcher, and Mic Mac, but varieties behave differently on different soils, so if you want to grow your own, a little experimenting is necessary.

We offer our berries by the quart at the roadside stand, or you can come out to the field and pick your own. Either way they're great, but remember—it's a short season.

Strawberry season is not complete without the old standby, Strawberry Shortcake. Modern times have changed people's tastes in this old favorite. The new sponge shortcake shells are a good example. What a way to ruin a strawberry!

A good shortcake is made with fresh, ripe strawberries, a good powder biscuit, butter, and heavy cream. On that, everyone agrees. Charlotte swears by one recipe, however, and Dick and Paula by another, so we've included both. You can't go wrong with either one!

DICK AND PAULA'S STRAWBERRY SHORTCAKE

4 cups flour	1 tsp. salt
4 tsp. Bakewell Cream*	½ cup shortening
2 tsp. baking soda	1½ cups milk

Mix and sift first four ingredients. Add shortening and mix with pastry blender. Add milk all at once and stir quickly with a fork to make a soft dough. Pour out on floured board and knead 5 or 6 times. Roll out ½ inch or more thick. Cut with biscuit cutter. Top each with a little milk or bit of butter. Bake at 475° for 5 minutes, then turn off heat and bake until golden.

Break apart while still warm and spread with butter. Cover the bottom with crushed, sweetened berries. Add the biscuit top, more berries, and a generous portion of heavy cream, not whipped or sweetened. Delicious for breakfast, lunch, or dinner.

*From the Bakewell Cream company comes the finest recipe for biscuits ever devised. Absolutely foolproof!

Berry picking time demanded an all-out effort from the family. As we trooped in from the strawberry beds on a hot June day it was heartening to visualize the juicy shortcake Aunt Ida would soon place before us. Here is how she did it.

CHARLOTTE'S STRAWBERRY SHORTCAKE

2 cups flour	1 quart strawberries
4 tsp. baking powder	hulled and chopped
4 tbsp. butter	¾ cup sugar mixed with berries
⅔ cup milk	and let set at room temperature
dash of salt	to gather juice

Sift together flour, baking powder, and salt. Lightly rub butter into flour mixture with fingers until coarsely crumbled. Gradually add milk, stirring after each addition until dough forms ball. Put into shallow cake pan, pushing dough out to edges. Bake 400° oven about 20 minutes. Remove from pan, split, butter, and cut into wedges. Add strawberries and serve hot with whipped cream or pouring cream. For 100 percent strawberry flavor prepare more berries and omit cream. This is the old-fashioned farm way.

Now, we want it clearly understood that this is in no way to be confused with Strawberry Shortcake. In fact, it is probably only because Polly is one of the authors of this book that she is even allowed to include a recipe which calls for a *cake mix*. However, this is one of the most elegant, scrumptious party desserts ever whipped up in no time flat, so here it is:

POLLY'S STRAWBERRY WHIPPED CREAM CAKE

Prepare a yellow cake from a good mix, substituting orange juice for milk or water. Divide into three round cake tins, and bake as directed, shortening slightly the time indicated for two layers. Cool on racks, and split each layer.

Whip 1½ pints of heavy cream until stiff, and sweeten lightly. Spread top layer of cake with a little of the whipped cream. Into the rest fold 1 quart of strawberries which have been hulled, sliced and sweetened to taste. Spread evenly over remaining five layers of cake and stack carefully, topping with the plain whipped cream. Decorate with whole strawberries. Cover tightly and refrigerate at least two hours. This will serve sixteen quite nicely. Even if you don't have sixteen people, you probably won't have many leftovers, but it will keep well for one or two days.

QUICK STRAWBERRY SHERBET

1 pint strawberries crushed   ½ tsp. gelatin softened in
½ cup sugar   1 tbsp. cold water
⅓ cup light corn syrup   2 tbsp. hot water
1 cup light cream

Let strawberries and sugar set together for a few minutes to gather juice. Stir in corn syrup. Dissolve gelatin in hot water and add to strawberry mixture. Lastly pour in light cream and beat all together in mixer bowl. Freeze in ice cube tray until about half frozen, then return to beater bowl and quickly whip it up again. Finish freezing and serve in crystal sherbet dishes.

Raspberries may be used in place of strawberries.

STRAWBERRY JELLY

3¾ cups strawberry juice   ¼ cup lemon juice
(about 3 quarts berries)   4½ cups sugar
1 bottle Certo, fruit pectin

Crush the fully ripe berries. Place in jelly bag and squeeze out juice. Measure strawberry juice and lemon juice into large kettle. Add exact amount of sugar, mixing well.

Place over high heat and bring to a boil stirring constantly. At once stir in Certo, then bring to full rolling boil and boil hard one minute, stirring constantly. Remove from heat, skim off any foam and pour quickly into sterilized jelly glasses. Cover at once with hot paraffin.

This is a very piquant breakfast jelly.

AUNT IDA'S STRAWBERRY PRESERVES

5 or 6 quart boxes strawberries picked with the sun on them
⅔ cup sugar for each quart of berries

Hull and rinse strawberries lightly. Put in good-sized kettle and add enough water to just cover berries. Measure and add sugar.

Select pint preserve jars, jar rubbers, and canning collar, and sterilize in boiling water. Be sure any utensil used is sterilized. Leave jars in the hot water to be filled.

Quickly bring fruit to a boil and cook rapidly for about 3 minutes or until berries soften slightly. Shut off heat and ladle into jars with rubbers already in place. Pop on covers and seal as each jar is filled. Cool gradually and store for winter in cool place. Do not expect these preserves to be the consistency of jam. They are whole berries floating in their own delicate syrup.

A patriotic Fourth of July celebration, c.1898. LMP

FOURTH OF JULY

Independence Day at Arrowhead in the nineteenth and early twentieth centuries meant a combination of public and private celebration. There was usually a trip to town to watch the annual parade and oration. Charlie always fondly reminisced about one occasion when the speaker was so inebriated that he could only mumble, "The Fourth of July! The grand and glorious Fourth of July! George Washington's Birthday! Hurrah for the Fourth of July!"

Another popular event was Fireman's Muster, where volunteer fire companies from surrounding towns met to compete at pumping water from their hand tubs. At home the old thirty-six star flag would be hung over the front door, and Charlie would bring out a tiny toy Civil War cannon which had been given to him as a child, produce a bit of gunpowder from his private closet, and fire it off. Later on, after dark, a few Roman candles and sparklers would be lighted on the lawn. In those days, when the strawberry crop was of much shorter duration, a familiar part of the Fourth of July was also scrounging in the strawberry bed for a last strawberry shortcake of the season.

Charlie on Dewey in Spanish-American War victory parade. LMP

These days, we've retired the little cannon out of respect for its age, and strawberries are usually still plentiful. But we still bring out the flag, although we've moved up to forty-eight stars, and there's still shortcake. Fourth of July Menu #1 is the traditional family meal, nowadays usually enjoyed on the piazza.

On an average day the farm stand may sell 200 quarts of strawberries and a dozen bushels of peas. But on the Holiday, demand for these Fourth of July favorites doubles or triples, so in the interest of providing the stand with fresh-picked supplies, Arrowhead workers usually must celebate the Fourth on the Fifth.

July Fourth menu #2 is the traditional cookout held for the farm crew.

FOURTH OF JULY MENU #1

Salmon - pan-fried w/cornmeal*, (or roast new lamb)*
Peas
Beets
New boiled potatoes
Parker House rolls*
Strawberry shortcake*
Pink lemonade*

Dip the fish in beaten whole egg and roll in cornmeal. Pop into a fairly hot fry pan in which a little butter has been melted. Lower heat immediately to low medium and cook slowly about 20 minutes to each side to produce a crispy golden meal coating and a sweet juicy interior. And do fry a little extra halibut or salmon for tomorrow's

FISH SALAD

Flake the cold fish into little chunks and place in a nest of crisped Big Boston lettuce. Pour over it a ladleful of homemade cooked salad dressing* and whisk to the luncheon table.

Baked stuffed fish seems to have infringed on the rights of pan-fried fish in recent years. But some of us still remember:
Tender fresh flounder filets sizzling in the pan
Inch-thick halibut steaks crisply golden outside, white and sweet inside
We remember tinker mackerel, our own catch from off Boar's Head
   or Plum Island, and how good they were right from the fry pan
Who can forget Fourth of July salmon steaks fried golden brown
   without and tender juicy pink within?

*Recipes included.

FOURTH OF JULY SUPPER

Vegetable salad w/new lettuce and cooked dressing*
Dish of strawberries w/cream
Bread and butter

SUMMER VEGETABLE SALAD

6 cups cold boiled new potatoes,
* chopped coarsely*
2½ cups cold cooked peas
1½ cups cooked beets, diced
* finer than potatoes*
2 tbsp. melted butter
salt and pepper to taste

Mix potatoes and peas together with seasoning. Stir in melted butter with fork, adding beets last and stirring until barely mixed to preserve color appeal of your salad. Refrigerate.

Serve in chilled lettuce cups with the following dressing:

COOKED SALAD DRESSING

2 cups milk
2 tsp. mustard
4 tbsp. sugar
2 tbsp. flour
2 tsp. salt
2 eggs, beaten
½ cup vinegar and
4 tbsp. butter boiled together

Scald milk in double boiler. Mix mustard, sugar, flour and salt thoroughly with spoon. Stir a little of the scalded milk into this mixture, then pour back into double boiler stirring constantly while it thickens. Cook 10 minutes. Stir in beaten eggs and cook not more than 2–3 minutes. Remove from heat. Add hot vinegar and butter. Strain, cool and store in refrigerator.

Be sure to use *sharp* vinegar, and don't let it boil away to scant measure.

PINK LEMONADE

juice of 2 lemons
1 pint strawberries or raspberries
about ½ cup sugar
5 cups cold water

Strain lemon juice. Mash berries and press through a sieve. Combine lemon and berry juice and stir in the sugar. Amount of sugar needed may vary according to tartness of juices. Pour into pitcher with cold water, add plenty of ice and head for the piazza.

JULY 4TH MENU #2

BARBECUE FOR THE FARM CREW

Hamburgers   Hot dogs
Lamb chops, or lamb leg sliced into steaks
Tenderloin steaks
German egg and potato salad*
Tossed salad   Macaroni salad
Watermelon
Chicken tarragon*
Onions and potatoes*
"Swamp Water" - Fourth of July Punch*

GERMAN EGG AND POTATO SALAD

5 lbs. potatoes	cider vinegar
8 hard boiled eggs	salt and pepper
3 large onions chopped	dry mustard
app. 1 cup mayonnaise	pinch garlic salt

Peel and cut potatoes, cook until soft. Chop eggs fine and add chopped onions. Add hot potatoes. Make dressing and mix together while hot. Chill thoroughly.

CHICKEN TARRAGON

Quartered chickens as needed	Herb salad dressing
Heavy duty foil	Tarragon flakes

Place chicken on foil, add generous amount of salad dressing and approximately ¼ tsp. tarragon. Seal tightly and cook approximately 45 minutes on grill. One packet per person.

POTATO AND ONIONS

New crop potatoes—2 medium per serving
Score tops, add 1 tbsp. butter and 1 tbsp. water
Wrap in foil and seal tightly. Cook 35-40 minutes. New white onions—done same way. Cook approximately 25-30 minutes.

SWAMP WATER—4th OF JULY PUNCH

1 qt. white rum
6-12 oz. cans beer

1 qt. 7-Up
2 small pink lemonade (frozen)

In large bowl put together in this order: 7-Up, rum, lemonade, and beer. Add ice and stir quickly. Be careful, this packs its own fireworks.

Charlotte, Paula, Justin and Dick on the piazza, July 4, 1980's. PCH

SUMMER SUPPERS—BREAD AND BUTTER AND FRUIT

"Bread is the staff of life" said our forefathers, and bread formed a major part of their diet. They knew the grains they grew contained the very elements of life.

Our forefathers also knew the fruits of the vine and tree were just as essential as bread. From time immemorial man has found them cooling and good at close of day.

Many a fine country supper has been planned basically around bread and butter and a dish of fruit in its seasons.

Let's begin with a proper loaf of bread. Many old New England breads used whole grains—wheat, rye, oats—or the unbolted "Graham flour" popularized by Sylvester Graham in the 1840s. But white bread was considered the finest, and became standard by the nineteenth century. This recipe of Aunt Ida's makes a firm, crusty loaf with a delicate taste—perfect as an accompaniment for summer fruits.

WHITE BREAD—AUNT IDA'S RECIPE

3 qts. flour   *2 cups warm water*
1½ tbsp. salt   *2 cups hot milk*
1 large spoon sugar   *1 or 2 yeast cakes*
               1 large spoon shortening

Dissolve sugar and shortening in hot milk, and yeast in ⅓ cup of the water. Sift flour and salt into large bowl. Add liquids and mix thoroughly. Turn onto floured board and knead until smooth and elastic. Return to bowl, cover, and let rise until double in bulk. Knead again, and separate into three loaves. Place each in a thoroughly greased bread pan, and allow to rise again until double. Bake at 400° 30 minutes. Reduce heat to 350° and bake 20 minutes more. (Aunt Ida's "large" spoon equals approximately 2 tablespoons.)

Wooden butter mold

MAKING BUTTER

    Butter making, long an important part of life on the farm had almost become a lost art in the twentieth century.

    For years our butter had been purchased at the butter and egg store, and good butter it was, too. Then along came World War II with its rationing program and butter lines. Now, farm people have all the patience in the world to wait for the weather to change or a crop to come in, but they thought very little of waiting in a long line of people to finally purchase a quarter part of what they wanted. Arrowheaders went home and brought out the butter churn and the milk pans.

The attic provided a variety of churns, including a large barrel type with crank handle—presumably the one Grandmother Ellen Ordway used to produce her prize-winning entries at the Amesbury Fair in the late nineteenth century. However, we chose a smaller one which, requiring less cream, was better suited to our present need.

This churn was a little upright one made of glazed earthenware and fitted out with wooden paddles attached to a long handle so they could be dashed up and down by hand. The milk pans were large round ones, made of tin, about 15″ in diameter, and shallow depth, 3″ or 4″.

When butter making was in process, several milk pans would be filled each evening with rich Guernsey milk out in the milk room. This room was like a little pantry and opened off the "back room" on the cool northwest side of the house. All milk equipment was kept there—milking pails, milk cans, measures, etc., and the milk was strained through a very fine cheesecloth as soon as it came in from the barn. The pine table, shelves, and floor in this room were scrubbed so clean they shone white.

In the morning the thick yellow cream which had risen to the top of the pans during the night was skimmed off by hand and set aside in a cold place until enough was gathered to fill the churn about two-thirds full. Then the cream was brought to ordinary room temperature and the churning started.

Unpredictable, the butter came sometimes fast and sometimes slowly. Up, down, up, down, went the paddles. Slollop, slollop, went the cream. When one began to think the butter never would come, suddenly the sloshing noise sounded heavier, and then, all at once, there it was—a soft mass, surrounded by whey about the consistency of water. The whey was drained off. The butter was lifted out and carefully washed with cold water.

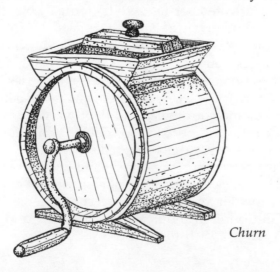

Churn

Salt, if desired, was added and worked in, about 1 tsp. to a pound, or it was left as sweet cream butter. Lastly, it was formed into pats or put into a butter mould to chill and harden for use. Butter came from the mould marked into squares with a pretty design in relief on each square. Guernsey milk, with its high butterfat content, makes the most delicious butter of all. Only heavy 40 per cent cream will make butter, so don't try to stretch it with light cream.

What of the byproducts? Well, the skimmed milk was fed to the hens, who were delighted, and the whey was washed down the drain. Today, we are told, whey is in great demand as an additive to many foods.

Although everyone enjoyed this delicious homemade butter, we were glad to return to the butter store after the war. Homemade butter is a treat, however, and if you don't have an old churn to experiment with, you can get acceptable results with your electric mixer or blender and heavy cream.

For supper, let's start with a big dish of sliced strawberries beside each plate. Sprinkle with sugar, but do not chill, as that changes the texture of the berry; and we'll fill the wild rose pitcher with light pouring cream. Now stack plenty of sliced bread on the wooden tray and bring out the butter dish. Then we'll finish with a lettuce cup on each plate filled with cottage cheese.

From strawberry time to raspberry time is only a hop and skip. Some fine evening try a large helping of sugared raspberries, chilled, and served with little hot buttered baking powder biscuits and thin slices of cold chicken.

In blueberry season most cooks prefer the wild ones for pies and muffins, keeping the large cultivated berries with their dusky bloom for table use. Sugar and chill these in advance. Now let's make a puffy omelet (as previously described) but add a sprinkling of mild cheddar cheese on top of omelet mixture after it is poured into pan. Cook as usual. The slightly zesty flavor is a good complement to the delicate blueberries. This time let's serve Syrian bread torn into pieces, buttered, and crisped in the oven. This may be done in advance of omelet and kept warm.

Last in sequence comes the peach crop, dainty white fleshed ones tinged with pink, or more robust-flavored yellow with rosy overtones. Some afternoon make a chowder of fresh corn and keep it warm. Select the peaches of your choice, peel and slice, then sugar and chill. At supper time heat up the chowder and add buttered common crackers. Place in a tureen on table and bring in the peaches, all yellow and pink, luxuriating in their own juice, a dish for each person.

Consider now the tomato—colorful, luscious, juicy. Slice several fully ripe beauties into a pressed glass serving dish and refrigerate. Slice the bread, buttering each one and place on glass platter. Bring out the silver sugar bowl and some silver serving forks for elegance. From the warming oven get the platter of little link sausages, panfried, brown and crisp, and retrieve the chilled tomatoes from the refrigerator.

Now circulate the food, letting each concoct his own bread and butter and sugared tomato masterpiece with sausage on the side.

FRUITS AND BERRIES

RASPBERRY TIME TREASURES

When the red raspberries are ripe, bake a Never Fail Sponge Cake.* Split and fill with 1 pint fresh raspberries crushed with 1/2 cup sugar. Pile top with whipped cream. This is a Summer Washington Pie.

When the purple raspberries are ripe, bake another Never Fail Sponge Cake. Chill and do not split this one. Pile on top a thick frosting made of 2 egg whites beaten very stiff with 1 cup sugar beaten in and 1 cup mashed purple raspberries folded into it. Serve immediately.

When the blackberries are ripe make a luscious blackberry pie, 2 crusts, if you please, and 1¾ cups sugar to a quart of berries for the filling, with just a dash of lemon juice.

FRESH RASPBERRY JAM

2 cups red or purple raspberries   1 cup sugar

Crush raspberries in bottom of saucepan and bring to boil very slowly over low heat. Stir in sugar and continue to simmer very slowly, stirring occasionally, for 20-25 minutes or until mixture thickens a little. Cool and use. That's all there is to it.

RASPBERRY TURNOVERS

In the pantry there are two covered wooden buckets, one for flour, the other for sugar. The first was once painted grey, but now the golden tones of old pine show through at points of wear. The other is all natural wood.

At least four generations of little feet have stood on these buckets to better see what was under construction when big folks brought out the pie board. Sometimes it might be cookies which would include ginger-bread men, or doughnuts with fascinating little holes to fry. In summer it was often raspberry turnovers, and an Arrowhead turnover is very distinctive, being pillow-shaped when finished and filled with freshly made raspberry jam. Usually, little folks could help make them.

The pastry for these turnovers is rolled out in a large thin sheet and cut into 5″ squares. Wet your forefinger with cold water and dampen all edges of the squares. In center of each one place about 2 tsp. freshly made raspberry jam, which has a brighter taste than jam from a jar. The bottom third of each square is now turned up over the jam and the top third is turned down so jam is completely covered. Press all edges just firmly enough to seal and turn ends back neatly against themselves to keep jam from escaping. Prick with fork 3 or 4 times and bake on cookie sheet in preheated oven 400° for 10 minutes, then lower temperature to 375°. They are done in about 15 minutes more or when crust turns creamy.

TAPIOCA WITH SUMMER FRUIT

Crush a pint of raspberries. Mix in 1/2 cup sugar and let stand together.

Measure 1/4 cup minute tapioca and 1/2 cup sugar into saucepan, blend together and stir in 2 cups water. Bring to a full boil, stirring constantly. Remove from heat at once. Tapioca clears as it cools. When cool, stir in raspberries and chill. Serve plain or with pouring cream. This is also good made with strawberries or peaches. For strawberries, increase berries to 3 cups. For peaches, use 2 cups, sliced small, and increase water by 1/2 cup.

BLUEBERRIES

Blueberries have been a New England favorite "ever since the Pilgrims landed." Even before the Pilgrims, in fact, the Indians had long enjoyed them; they had cleared large areas by burning so that blueberries, raspberries, and strawberries could grow more abundantly. As a child of seven or eight, Anne Pollard was the first person off the boat when the Puritans landed to found Boston. Asked at the age of 100 what the peninsula had looked like then, she remembered only that it was covered with blueberry bushes. John Josselyn, who visited New England in the mid-seventeenth century, was enchanted with its "Skyecoloured berries." "They usually eat of them put into a Bason, with Milk and sweeted a little more with Sugar and Spice," he said, adding that they made "a most excellent summer dish." They still do. Cultivated blueberries were introduced in 1913, and are delicious for eating, but wild ones are still best for cooking.

WILD BLUEBERRY PIE

3 cups wild blueberries　　　　　*1¾ cups sugar*
4 tbsp. water

Roll out your favorite pastry mixture and line shallow 9″ pie plate. Rinse berries and put into plate. Dust over very lightly with sifted flour, cover with the sugar, and add the water. Seal with top crust. Pop into preheated 450° oven. Bake 10 minutes, lower heat to 350°, and bake until crust is done and pie is bubbly in center slit.

BLUEBERRY CAKE

3 tbsp. butter　　　　　　　*1¾ cups flour*
1 cup sugar　　　　　　　　*2 tsp. baking powder*
1 egg　　　　　　　　　　　*¾ cup milk*
1½ cups blueberries, lightly floured

Cream shortening; add sugar, beaten egg and milk; add flour sifted with baking powder and stir in blueberries. Bake in greased shallow 11″ pan in 375° oven. Serve hot with butter. *If* there's any left over, this is also good cold as a dessert or snack.

CHARLOTTE'S BLUEBERRY MUFFINS

1 cup blueberries　　　　　　*1 tsp. salt*
3 tbsp. sugar　　　　　　　　*2 tbsp. sugar*
2 cups flour　　　　　　　　　*1 egg, beaten well*
3 tsp. baking powder　　　　　*1 cup milk*
6 tbsp. melted shortening

Sift flour, baking powder, salt, and 2 tbsp. sugar into mixing bowl. Make a well in center of flour and add liquid—egg, milk, and melted shortening—all at once. Stir until just mixed and still lumpy. Add berries. Stir in quickly. Fill muffin tins 2/3 full. Sprinkle with the 3 tbsp. granulated sugar. Bake 25 minutes. Oven 425°.

BLUEBERRY JAM

Crush wild blueberries to equal 4 cups (about 3 pints). Add 2 tbsp. lemon juice. Place in large (6-8 qt.) heavy saucepan. Add 1¾ oz. dry fruit pectin. Bring to full boil. Add 4 cups sugar. Bring to full boil for one minute. Remove from heat, skim off foam. Pour into scalded glasses. Pour hot paraffin on top or use two-piece metal lids and process 5 minutes in a water bath canner.

SUMMER APPLE PIE

Make pastry for two-crust pie. Select six or seven summer apples, tart, but not so green they are sour. Peel apples and slice into a shallow 9" plate (about 3 cups fruit). Cover with 1 3/4 cups sugar and a fine sprinkling of flour. Add 3-4 tbsp. water. Season with a little fresh grated nutmeg only. This enhances the delicate flavor of summer apples, and does not overpower it. Bake as previously described.

PEACH PIE

Prepare enough of your choicest pastry for a two-crust pie.

Into the lined 9" pie plate place enough thinly sliced ripe peaches to make a rather thin filling. This is not a deep dish pie, and there should be a nice balance between crust and filling. Lightly dust on a sprinkling of flour with sifter and cover with 1 3/4 cups sugar. Now distribute about 1 3/4 cups water over the sugar and seal on top crust. Put into preheated 450° oven. After 10 minutes lower heat to 375° and bake till center slit is bubbly.

This is a nice way to use peach "drops" or second-size fruit. But remember you will not get a delicious pie from unripe or poor flavor fruit.

Country people consider a proper peach pie one of the true delights of summer.

PEACH SHORTCAKE

An old favorite, often neglected these days. Pity! Sweet, juicy, fully ripe peaches make a shortcake every bit as luscious as its better-known cousin, the strawberry shortcake. Proceed as for strawberry shortcake.

PEACH COBBLER

Make a shortcake mixture as described under strawberry shortcake. Set aside about 1/3 of the dough. Roll out the rest to 1/4" thickness and line a rectangular 9" or 10" cake tin pulling dough up against sides and end of pan. Fill with

3 cups sliced peaches	¾ cup sugar sprinkled
¼ cup water added	over them

Roll and cut remaining dough into strips and put a widely spaced lattice top on peaches. Bake at 400° till peach is bubbly and crust golden (about 25 minutes).

Serve hot with pouring cream or whipped cream.

TO PRESERVE PEACHES AND PEARS

Do put up for winter a few jars of these later summer fruits. Peel your peaches or pears and cut into either halves or quarters. Two-thirds cup of sugar to a quart of fruit is a pretty safe rule, but you may want to vary it a bit as fruit varieties vary some in acidity and sweetness.

Follow preserving instructions for strawberries, remembering these firmer fruits require a little longer cooking. But watch them! Test them with a fork to see when they are done to perfection. Then quickly fill jars and seal.

PEACH CONSERVE

3 pints chopped-up peaches	1 cup seedless raisins
2 large oranges cut up	5 cups sugar
3 cups water	

Cook very slowly (1 1/2 to 2 hours) until it thickens and begins to look glassy. Then add 1 cup chopped walnut meats and seal in pint preserve jars. May also be stored in glasses with paraffin seal.

Charlie contemplates modern improvement, c. 1947. CMC

COLD MEATS FOR SUMMER SUPPERS

Make these ahead, preferably in the cool of early morning, and enjoy several delightfully simple meals.

PRESSED LAMB

Buy a forequarter of spring lamb, and have it cut into several pieces. Place meat in kettle and barely cover it with water. Boil slowly with kettle covered until meat is tender and comes away from the bone easily. Remove any excess fat, but be sure to leave some of the fat as this helps set up the loaf.

Press meat through grinder while still hot. Salt the meat a bit and run pepper mill over it lightly. Meat should be rather moist. If it seems too dry, add a little stock from the kettle. Now pack it rather firmly into buttered mold, glass or pottery, cover with foil or wax paper and press down firmly with weight while it cools. Refrigerate.

Sliced cold this makes a memorable supper with hot buttery peas, biscuits, and a dish of sliced strawberries with pouring cream.

COLD BOILED TONGUE

One of the most neglected meats nowadays is fresh beef tongue. Country folk have always known that it is a great delicacy, especially in summer, when cold sliced meats make such a nice accompaniment to all those delicious vegetables. Fresh tongue is sometimes available in supermarkets, but you may have to ask your butcher to order it for you.

Put a 3- to 4-pound fresh tongue in a large kettle and cover with cold water. Add 1 medium onion, quartered, 1 stalk of celery with its leaves, 1 carrot, 4 or 5 peppercorns, and a pinch of salt. Bring to a boil and simmer for 3 hours, or until the tongue is tender. Let it cool in its broth. When it is cool enough to handle, peel off the skin and trim off the root end. Place tongue in a bowl or other deep container and strain broth over it. Cover and chill. Slice as needed, being sure each time to include some meat from both ends of tongue, as the texture differs significantly, and most people prefer some of both.

JELLIED CHICKEN

Place cut up chicken or fowl in kettle with 1 onion sliced and 1 stalk celery with leaves. Cover with water, cook very slowly until meat falls away from bones easily. Remove chicken from stock, skim off fat and boil down liquid until you have 2 cups. Then add 1 tbsp. gelatine which has been soaked in 4 tbsp. water. In bottom of a glass mold place a few sprigs parsley and 1 or 2 thin slices sweet pepper. Pack meat into mold in small pieces, sprinkle layers with little salt and pepper, a few cooked peas and finely chopped bits of celery. Pour hot stock over it, place weight on top of mold, and store in refrigerator till very firm.

This is an excellent dish for a hot summer evening.

A quick and tasty hot supper, especially nice when the new peas are in season.

CREAMED BEEF

4 tbsp. butter	*2 cups milk*
4 tbsp. flour	*salt and pepper*
1 cup cooked peas	*1 jar thinly sliced dried beef.*

Make a thin white sauce as follows: melt butter; add flour, seasoning, and milk, stirring until it begins to boil. Continue to boil for 5 minutes, stirring constantly. Add: peas and dried beef. Serve on toast.

SUMMER VEGETABLES

AUNT IDA'S VEGETABLE MELANGE

Cook until tender:

½ cup diced carrots	*2 cups milk*
½ cup peas	*2 tsp. flour*
½ cup bits of celery	*2 tsp. butter*
¼ cup white turnip	*Dash of salt*
¼ cup yellow snap beans	

Aunt Ida had a way of coming up with tasty little hot dishes to perk up a meal which might be a bit drab otherwise. When she prepared vegetables she often cooked enough to have some left over, and she used them.

She would make a very, very thin white sauce by blending flour into milk and adding butter, stirring while it cooked and thickened. Then she'd get out her ready cooked vegetables: carrots and/or peas (lacking peas she would sometimes cook celery bits till just soft). Maybe she'd find a little white turnip to slice, or a few yellow snap beans.

When she finished, there would be the best little side dish of hot buttery milk with assorted vegetables you could ask for.

CANDIED CARROTS

Cut a bunch of carrots (6 or 8 small ones) into sticks. Place in saucepan, little more than covering with water. Add 3 or 4 tbsp. brown sugar and 2 tbsp. butter. Cook slowly, adding more water if necessary, for about 20 minutes. Then boil down water (carefully) till carrots look glazed. Taste. They should be a little firm, a little sweet, and a bit buttery.

SAUTEED ONIONS

Select 5 or 6 medium small onions for finer flavor and texture. Slice into frypan with 2 tbsp. melted butter. Cook slowly over medium low heat about 20 minutes, turning occasionally, but not too often as you want them to brown a little. When golden and still a bit firm put them on hold until your steak or broiled hamburg is ready.

SUMMER SQUASHES

No garden should be without summer squashes. They are versatile, productive, and require little care or space. Yellow "crooknecks" and the delicate white "Patty" squash are staples, but the most popular is the zucchini. On any given day, five to fifty tons of this slender green squash pass through the Boston Market, and who knows how much lurks in the backyards of suburbia?!

It happens every year! Zucchini are cut for market every other day. Some grow too large in 48 hours, and the pickers have inevitably missed a few hidden among the foliage at the previous cutting. These must be picked, and discarded, for the plants to continue bearing.

New people on the crew, appalled by this waste, carefully set the large squash aside to take home for zucchini bread. After a few hours, however, they realize that a few hundred pounds of squash a day will make a little bit too much bread!

ZUCCHINI BREAD

3 eggs
2 cups white sugar
1 cup salad oil
2 cups grated zucchini
3 tsp. vanilla

1 cup chopped walnuts
3 cups flour
1 tsp. salt
1 tsp. baking soda
¼ tsp. baking powder

3 tsp. ground cinnamon

Beat eggs until light colored and foamy. Add sugar, oil, zucchini, and vanilla. Mix lightly. Combine dry ingredients and add to egg mixture. Stir until well blended. Add nuts. Pour into two 9" x 5" loaf pans. Bake in preheated 350° oven 1 hour. Cool on racks.

HONEY-NUT ZUCCHINI BREAD

3 eggs
1½ cups honey
1 cup butter
2 cups grated zucchini
4 cups whole wheat flour

3 tsp. vanilla
1 tsp. baking soda
¼ tsp. baking powder
3 tsp. cinnamon
½ cup chopped nuts

1 tsp. salt

Combine eggs, honey, butter, salt, zucchini, and vanilla. Mix dry ingredients together and add to egg mixture. Pour into 2 buttered loaf pans. Bake 1 hr. at 350°. Best if refrigerated before eating.

 *Polly watches her grandfather haying, 1940's (left)* CMC;
Dick and his corn pickers (above);
Justin enjoys haying as much as earlier generations did. CDC

ZUCCHINI PARMESAN

In large skillet, melt 2 tbsp. butter with 2 tbsp. vegetable oil. Add sliced zucchini (1 small zuke per serving). Sprinkle generously with garlic powder. Cover and cook over medium-low heat, stirring occasionally, for 20 minutes. Add grated parmesan to taste, cover, and let cheese melt.

POLLY'S ZUCCHINI-CHEESE SOUP

3–4 lbs. zucchini, sliced	2 cans cream of chicken soup
3 stalks celery, sliced	2 soup cans milk
5 green onions, chopped	½ cup sharp cheddar cheese,
including green part	grated
water to cover vegetables	salt
2–3 chicken boullion cubes	fresh ground pepper

1 cup light cream

This is a recent addition to the Arrowhead repertoire—as is zucchini, which Dick introduced when he started farming. It's good hot in the fall, but it's best in the summer chilled.

Simmer vegetables in water with bouillion cubes until tender. Puree in blender; return to kettle. Add chicken soup and milk and blend thoroughly. Heat to just below boiling point. Stir in cheese, and season to taste with salt and pepper. Serve hot or cold, stirring in light cream just before serving. Garnish with chopped fresh parsley, if desired.

SAUTEED SUMMER VEGETABLES

For each person you will need:

1 small zucchini   1 small tomato   1 small summer squash

Slice squash thinly and place in frypan in which you have melted a little butter. Sautée slowly until they begin to brown lightly. Peel tomatoes and slice into pan. Season to taste with salt, freshly ground pepper, minced garlic, oregano, and a tiny bit of thyme. Cover and cook over low heat. As liquid evaporates, add small amounts of water to prevent burning and keep it juicy. When squash is tender, reduce heat and keep warm until serving. A little chopped pepperoni or Genoa salami added with the tomato makes this a tasty one-dish meal.

CUCUMBER PICKLES; RELISH

Pickles should be made in the hot, humid days of July, when cukes are plentiful; not in the fall when the pickling mood strikes the cook. In the cool weather of fall, cukes are in short supply and often expensive. We can recall a few fall forays to the pickle patch when the quality and quantity of the crop were disappointing and we had to go without pickles. The moral is: if it's too hot to think about pickling, it must be the right time to do it.

BREAD AND BUTTER SLICES

4 qts. sliced unpared
  pickling cucumbers
6 medium white onions,
  sliced (6 cups)
2 green peppers,
  sliced (1⅔ cups)
3 cloves garlic (whole)

⅓ cup granulated
  pickling salt
5 cups sugar
3 cups cider vinegar
1 ½ tsp. tumeric
1 ½ tsp. celery seed
2 tbsp. mustard seed

Combine vegetables and garlic. Add salt, cover with cracked ice, mix thoroughly. Let stand 3 hrs., drain well. Remove garlic. Combine remaining ingredients, pour over vegetables. Bring to boil. Fill jars and process 5 minutes in water bath canner.

VEGETABLE RELISH

7 large yellow onions
1 medium head cabbage

10 green tomatoes
12 green sweet peppers

6 red sweet peppers

Using coarse blade on grinder, put all vegetables through without cleaning between vegetables into a large kettle. Sprinkle with pickling salt and let stand overnight. Rinse and drain.
    Combine remaining ingredients:

6 cups white sugar
2 tbsp. mustard seed
1 tbsp. celery seed

1½ tsp. tumeric
4 cups cider vinegar
2 cups water

Pour over vegetables, boil gently 5 minutes. Fill jars and process 5 minutes in water bath canner.

TOMATOES

On the wholesale market, tomatoes are a hit-or-miss crop, red gold if they're early, red ink if they're not. But we find our retail customers will gladly pay a premium price for top quality field-ripened tomatoes all season.

We start our transplants in the greenhouses in March, and set them in the field in mid-May. This early crop is grown on plastic mulch, which hastens their growth, so we usually start picking in early July. A later planting is also made to extend the crop through October.

The most popular ones are the large, firm red varieties — Starfire, Jetstar, Supersonic, and Beefsteak — in order of maturity, but we also offer Basket Pack, a real nice cherry-fruited type, and Roma, for paste and sauce.

TOMATO CASSEROLE

This is a version of one that Aunt Ida used to make.

Butter well a quart glass bread pan.

Butter 4–6 slices (depending on size) white bread (remove harder top crust) — then cut or break into pieces about 1"–2" each square.

Line the bottom with one layer — saving rest.

Wash (remove stem and skin if desired) of 2 very large tomatoes or 3-4-5-6 smaller tomatoes according to size. Cut into slices and spread some over the bread — another layer of buttered bread pieces — more tomatoes, etc., topping with bread. Between each layer a dash of salt and pepper and 1/2 tsp. to 1 tsp. sugar (more if liked sweeter and also depending on the ripeness of tomatoes). If tomatoes are not really juicy sprinkle a spare amount of water over the 2nd and 3rd layers near the top. Dot well with good chunks of butter and sprinkle on a bit more sugar. Bake about 375° 35–40 minutes. Serve hot.

"Well, I hope you're going to have something for me to do with green tomatoes in September!" said a friend when we mentioned writing this book. No problem. We could no more fail to include Arrowhead Farm Tomato Relish in this book, than we could fail to make it every fall. Saturday night baked beans don't taste *really* right without it; hamburgers and hot dogs are improved by it; and it makes the best base for Meat Pudding. Like pickles, however, it *should* be made in the summer, when both red and green tomatoes are at their peak.

TOMATO RELISH

2 qts. ripe tomatoes, coarsely chopped
2 qts. green tomatoes, coarsely chopped
8 medium yellow onions, peeled and coarsely chopped

Sprinkle 1/4 cup salt over tomatoes. Let stand 24 hours; then rinse and drain.
Make syrup of:

1 qt. water   *1 tsp. cinnamon*
1 qt. cider vinegar   *1 tsp. celery salt*
2 lbs. brown sugar   *1 tsp. dry mustard*
1 tsp. whole cloves, tied in cloth

Boil 5 minutes. Add tomatoes and onions. Simmer slowly uncovered, for three hours. Ladle into sterilized jars and seal.

HOMEMADE TOMATO SOUP

1 scant peck ripe tomatoes   *2 medium sweet peppers*
7 stalks celery with leaves   *5 medium onions*
2 small hot peppers   *25 whole cloves in small cloth bag*

Immerse tomatoes briefly in boiling water and remove skins. Cut into chunks and mix with cut-up vegetables. Be sure to remove all seeds from peppers. Place in large covered kettle and put in the bag of cloves. Cook very slowly until soft. Remove from heat. Take out spice bag.

Press mixture through a potato masher, and then rub through a sieve. Then add:

2 cups sugar   *2 tbsp. salt*
1 cup flour

which have been thoroughly mixed with spoon. Cook together slowly until mixture thickens, stirring very frequently so it can't adhere to bottom of kettle. When thickened and simmering, remove from heat and seal in pint jars as described in preserved strawberries. Dilute one-half when served, and add butter. Plain bread croutons are delicious with this.

Actually, the aroma of this soup simmering in your kitchen is almost enough reward for making it.

This recipe may also be adapted to a blender. Because the blender will result in more of the vegetable pulp being retained, you will want to increase hot peppers to 2 1/2–3, and cloves to 35–40. Allow mixture to cool slightly after removing spice bag, then blend at medium speed until smooth. Rub through sieve and proceed as above.

What Yankees used to call chili sauce, and what is now called chili sauce, are two totally different things; and where Yankee chili sauce got its name is a mystery, for most of the many recipes we've seen have absolutely no hint of chili peppers in them. In fact, they seem to be more or less interchangeable with recipes for catsup, of which we also have many. This one, which is the one most often used within our memory, appears in Aunt Ida's hand as "Shirley Sauce," and perhaps that is the best name for it. It was often served with yellow eye baked beans on a cold Saturday night with steaming brown bread and butter.

AUNT IDA'S "SHIRLEY" SAUCE ("CHILI" SAUCE)

6 large ripe tomatoes	*2 tbsp. sugar, light brown*
2 small red peppers, seeds removed	*1 tbsp. salt*
2 medium onions	*1 cup sharp vinegar*

Chop tomatoes, peppers, and onions fine. Combine with other ingredients and cook slowly for one hour. Put through a sieve and pour into little hot sterilized jars and seal with paraffin.

AUNT IDA'S WORCESTERSHIRE SAUCE

25 ripe tomatoes, chopped	*2 tsp. salt*
10 green peppers,	*1 tsp. cloves*
seeds removed, chopped	*1 tsp. nutmeg*
4 onions, chopped fine	*1 tsp. allspice*
½ cup sugar	*1 qt. vinegar*

Simmer 2 hours or more until thick. Bottle hot and seal with paraffin.

As with chili sauce, this recipe bears little resemblance to the Worcestershire sauce sold in stores. We assume that the name, which is commonly used for this sort of tomato-pepper-spice concoction, was adopted because of its spicy connotation.

AUNT RUTH'S TOMATO CATSUP

1 peck tomatoes	1 tbsp. cloves tied into
(to make 1 gallon strained)	little bag
6 tbsp. salt	2 tbsp. cinnamon
3 tbsp. black pepper	2 tbsp. allspice

1½ pts. vinegar

Cook tomatoes 20 minutes, then strain. Add spices and vinegar and boil down to half quantity. Seal in sterilized jars and enjoy with baked beans.

BEANS

In the 1900s dry beans for baking were an important crop on the farm. Being plants tolerant of poorer conditions, they were right at home on a piece of sandy land at the southeast corner of the farm. A special strain of yellow eyes, developed on the farm, was planted in this field each June. After being processed in the fall, they were hauled to the B & M Railroad Depot in Newburyport, and shipped to the cannery in Portland. Unfortunately, this piece of land was sold in the 1940s, and is now a housing development. We must admit, however, that these same houses provide a pool of young people who are eager for summer picking jobs.

Today, our crop is limited to green and yellow snap beans. Weekly plantings, from late April through July, provide young tender beans for our customers all summer and fall. The new varieties, bred for machine harvest, have lost some of that true "beany" flavor; so we still grow the old standbys—not as prolific, but really good eating.

Pick your beans while the water is coming to a boil, remove the ends, snap into one-inch lengths, cook till just tender, butter and salt to taste. A meal in itself.

CORN

Every year around Patriot's Day, when the soil temperature is approaching 50°, the first planting of sweet corn goes in. A new piece is planted every four or five days, until early July. This continuous planting assures us of a steady supply of young tender corn later in the year, with a new field ready to pick every few days. Occasionally the weather interferes with the succession, but it usually works out fairly well.

Early in the season we plant varieties more tolerant of cold, wet soil (Spring Gold, Sprite, and Harmony). As the soil warms in May, we switch to better quality main crop varieties, such as Sweet Sue, Sweet Sal, Symphony, and Silver Queen. Most of these corns are bi-colors or butter-and-sugar types, which are the most popular.

Spring Gold is yellow but earlier by a week than the others, and is well received when the first picking starts in mid-July. Silver Queen, an all white variety, is the last to ripen. Its quality and flavor are unsurpassed, so we make several plantings to mature from late August through October.

About 25 acres of sweet corn are grown on the farm, and since we retail the whole crop, our only concern is quality. So we grow only the best varieties and pick our corn fresh several times each day. Once an ear of corn is removed from the plant its sugar quickly converts to starch. So put the water on to boil before you drive out to the farm. With sweet corn, freshness is everything.

If you'd like to grow your own, great! For good pollination, corn should be planted in blocks, not rows, about 18" apart each way. It's a heavy feeder, so don't spare the fertilizer, especially nitrogen. Ample water, a full-sun location, and good weed control round out the program. It's a rewarding crop, but remember—only one ear to a stalk so plant plenty!

CORN CHOWDER

6-8 medium ears of corn	*4 tbsp. butter*
5-6 medium size potatoes, sliced	*5 cups milk*
1 medium onion, sliced	*salt and pepper to taste*
8-10 common crackers split	

Cook fresh corn in covered kettle with little water about 12 minutes. Remove corn and add onion and potatoes to corn water. Cook slowly until just soft. Slice corn from cob with sharp knife and add, together with the milk and butter. Bring to a boil and simmer five minutes. Salt and pepper to your taste. Hold on warm burner for 30-60 minutes without further boiling to gather flavor. Five minutes before serving add the crackers which have been soaked for 20 minutes in a little cool milk.

Two cans cream style corn may be used instead of corn on the cob. Either way they will want seconds.

CORN IN MILK

Slice cooked corn off cobs into saucepan with very sharp knife. Add milk to cover. Heat slowly until it begins to simmer. Remove from heat. Add butter and salt to taste. Enjoy your breakfast.

SUMMER DESSERTS

Every Sunday in summer when Charlotte and Elizabeth were growing up, we had ice cream for dinner dessert. We froze the cream in a hand-cranked freezer packed with ice and rock salt. (When we ran out of this, we substituted coarse-fine hay salt from the barn.) We set up this ice cream operation in the woodshed where the tub could drain harmlessly as the ice melted. The freezing usually took 20 to 25 minutes. A patient father turned the gears while the impatient children waited for the cream to freeze. Then the dasher would be removed, all dripping and rich, and they could take turns licking it.

SARAH'S ICE CREAM

1 qt. milk   *1 cup sugar*
3 eggs   *½ cup flour*
                pinch of salt

Scald milk and set aside to cool. Separate eggs; whip whites stiff; beat in yolks one at a time. Mix together sugar, flour, salt.
Combine this together with egg mixture in large bowl and mix thoroughly. Heat 2 extra cups milk hot and pour over mixture, stir all together and put into double boiler. Cook 15 or 20 minutes. Remove from heat, add flavoring, 1 tbsp. vanilla, and pour into large bowl or heavy pitcher. Now add the quart of scalded milk and stir in thoroughly. Cool overnight. Just before freezing, whip a cup of heavy cream and fold in. Freeze and eat with chocolate sauce.

CHOCOLATE ICE CREAM

Omit the vanilla. Melt 3 squares baking chocolate over water and stir in 1/2 cup sugar. Add to the custard mixture and freeze.

STRAWBERRY ICE CREAM

Omit the vanilla. Chop 2 quarts strawberries and stir in 1 cup sugar. Add to custard mixture and stir in. Freeze.

PEACH ICE CREAM

Omit the vanilla. Chop 10-12 peeled peaches and stir in 1 cup sugar. Stir into custard mixture and freeze.

CHOCOLATE SAUCE

2 cups sugar
4 tbsp. cocoa (or 4 tbsp.
  carob powder)

1 cup water
2 tsp. butter
1 tsp. vanilla

Mix sugar and cocoa thoroughly in saucepan. Stir in one cup water, add butter and bring to a boil, simmering slowly 15-20 minutes. Remove from heat. Add vanilla. Serve over vanilla ice cream.

"CHARLOTTE ROUCHE"

A country version (and pronunciation) of Charlotte Russe, this was always a favorite with summer boarders and home folks alike.

2 cups heavy cream
⅔ cup powdered sugar
3 egg whites beaten stiff

1 tsp. vanilla and
2 squares chocolate melted
  over boiling water

Beat cream until stiff. Add sugar, fold in egg whites and flavoring. Line individual sherbet stemware with narrow strips of sponge cake or lady fingers. Fill with mixture. Chill at least 1/2 hour before serving.

CHOCOLATE CAKE WITH BOILED FROSTING

2 squares chocolate
½ cup milk
1 egg yolk
1 tbsp. butter

1 cup sugar
1½ cups sifted flour
1 tsp. soda
½ cup milk

1 tsp. vanilla

Shave chocolate thin and cook with milk and egg yolk until it boils. Remove from stove and add butter and sugar. Pour mixture into flour; add soda, dissolved in milk, and vanilla. Pour into 8" sqare pan and bake in oven at 375°.

BOILED FROSTING

1 cup sugar   *3 tbsp. water*

1 egg white

Boil sugar and water together until it will spin a thread. Remove from stove and slowly pour it into stiffly beaten egg white with beater set at high speed. When mixture begins to stiffen somewhat, quickly spread it over the cooled chococlate cake.

PICNICS

We don't know exactly how long picnicking has been a part of summer at Arrowhead, but we suspect for a long, long time. We do know that at least as early as the end of the eighteenth century, people from this area went on day-long berrying and fishing expeditions to Plum Island. And Aunt Lizzie's photographs from the 1890s include many picnic scenes.

While we haven't given up picnicking close to home, the twentieth century introduction of the automobile has broadened the range of picnicking. Glen was perhaps the most indefatigable picnicker of us all, and expanded our vision of the picnic season. Within the memory of the current generation, the earliest picnic took place on an unusually warm day in February, on the rocks at Magnolia. And the latest was the day after Thanksgiving at Reed State Park on the coast of Maine, when we all ate our sandwiches with mittens on.

Herewith, some of our favorite summertime picnics:

A HORSE AND BUGGY BEACH PICNIC
IN THE 'TEENS

Picnics have a happy way of lingering in one's earliest memories. A prime event of the summer at Arrowhead in the 'Teens was a trip to the seashore. On a sunny August day we hitched our horses into the two buggies, stowed their feed bags under the seat, packed in our own lunch baskets and the field glasses, and headed for the beach through the old "Rabbit Road" short cut. The horses' hoofs picked up little puffs of sandy dust from the dry road between the high bush blueberries on either side. Near the small plank bridge over a clear brook, a cart path led down to the water. Our horses liked to stop here for a cooling drink while Father provided each carriage with several branches of blueberries for refreshment as we rode along. Then we drove through the brook and rejoined the road.

The beach was beautiful. In those years the seaside birds still nested among the beach peas and grasses. We unhitched the horses, tethering them to a wheel with a rein long enough so they could roll in the sand before they enjoyed their oats. The older folks spread down a robe on the sand and opened up the lunch, while the children raced to the ocean. Arrowhead children of this era never went barefoot at home, nor did they want to; however, the fine sand felt delightful to bare feet and so did the rolling and ebbing salt water.

But the lunch called: sandwiches of cold sliced ham, home-boiled, and of egg salad, a dish of sliced peaches with sugar, and, "brambles" with their delicious raisin and lemon filling. The thermos contained lemonade. After eating, each relaxed in his own fashion: digging in the sand, wading in the ocean, collecting shells, or merely sitting in the sun.

All too soon we were travelling back. When we turned into our home road the horses quickened their pace. All had enjoyed the day—all were glad to be home.

Sarah, Elizabeth, and the horse enjoy the beach, c.1912. LMP

BRAMBLES

Line an 8″ rectangular pan with *piecrust*. Make a filling of:

1 cup seeded, chopped raisins   1 cup sugar
1 lemon; juiced and grated   2 tbsp. flour stirred into
1 egg, beaten   ½ cup water
little salt

Mix all together and cook 10 minutes in saucepan. Pour into piecrust, top with another sheet of pastry. Seal edges and prick top. Bake at 400° until pie crust is golden. Cut in squares.

HONEY OATMEAL COOKIES

½ cup vegetable oil   1 cup whole wheat flour
2 eggs   ¾ tsp. baking soda
¾ tsp. salt   3 cups quick rolled oats
¾ cup honey   ½ cup chocolate bits
1 tsp. vanilla   ½ cup raisins

Beat oil, eggs, salt, honey, vanilla until smooth. Add flour and baking soda. Stir in oats and chocolate and raisins. Drop by teaspoons onto greased sheets. Bake 10-15 minutes in 350° oven.

TAKING THE ELECTRICS TO YORK BEACH — 1920

Although the electric cars were a blessing to many travellers, we seldom used them because the line was so remote from our farm. However, one delightful summer day we "took the electrics" to York Beach. The whole family went down the road on foot, each carrying a share of the picnic duffle. We took the nearest route through the pines along the river about a mile, making good connection with our conveyance.

This was a summer car, open on all sides, and it hummed merrily as it gathered speed and rushed through the countryside. Tree branches and wild flowers along the roadside bowed and danced in the wake of wind created by our passage — a ride to charm any child.

We arrived at York in plenty of time to play and relax by the ocean before lunch. There were sandwiches of thinly sliced roast beef with its own gravy, tiny pickles and saltine crackers to munch with cheddar cheese (Father had tinned sardines with his crackers), and then out came the box of plump little pillow-shaped turnovers filled with fresh strawberry jam.

Glen and Polly picnic along the Mohawk Trail, 1940's. CMC

There were also Monkey Face cookies and a thermos bottle of hot tea—a feast to remember.

After lunch in the fresh salt air, we strolled through the town: Mother to visit the famous linen and rug shops, the children to look at the brightly colored toys and gaudy gift items in what Father termed the "junk shops." On the ride home, Mother and Aunties treasured a new drawn-work linen tablecloth with napkins to match. Father had purchased Goldenrod kisses. The children had little boxes of compressed paper nuggets from Japan, which blossomed into tiny exquisite flowers when floated on water.

A happy trip. But somehow, the path home through the pines had grown much longer.

MONKEY FACES

1 egg	*a small tsp. soda*
1 cup sugar	*2 tbsp. butter*
½ cup sour milk	*2 scant cups flour*

Cream together butter and sugar and add beaten egg. Dissolve scant tsp. soda in sour milk and combine with creamed mixture. Gradually stir in flour. Drop batter by small spoonfuls on buttered cookie tin. Into each cookie press three raisins: two for eyes, one for mouth. Bake 375°.

A PATRICIAN WHITE MOUNTAINS EXPEDITION–1950s

After World War II years, when there had been no unnecessary travelling, it was a treat to go further afield again. Our vehicle in the 1950s was a Patrician model Packard. Big enough for oldsters and youngsters all to pile into, "Patricia" was also a wonderful car to drive, although Glen grumbled about getting "ten gallons to the mile."

The huge rear seat had a fold-down arm rest, which the youngest child used as a perch to better view the world, and the enormous trunk held enough gear for this generation of practically professional picnickers: an ice cooler, basket of picnic equipment, wool blankets, extra jackets, bathing suits, camera, binoculars, burlap bags, heavy twine, a few basic tools, a trowel, and some plastic specimen bags—who knew what gem you might encounter en route? We made frequent stops to appreciate roadside nature, or to relieve the pressure of excess energy in the rear seat.

Eventually reaching Lake Chocorua, we would go for a swim, afterwards watching the loons with binoculars and trying to capture the clear reflection of the mountains in the lake on film. Then we ate our lunch: crabmeat or tuna salad sandwiches, stuffed eggs, with a dish of sugared raspberries, and some potato chips washed down with cold Coke, and finished up with marshmallow filled chocolate drop cakes and cold milk. After that one child got out a sketching pad, the other hunted for frogs with Mother. Soon Father returned from a short walk and took us all to see signs of beaver he had found.

After a perusal of the balsam pillows and tiny birchbark canoes in the gift shops and a sundae or ice cream soda in Conway, we hit the road home, this final leg of the journey inevitably bringing a reference to Norman Rockwell's Saturday Evening Post cover of the tired picnickers returning home.

Nothing makes a picnic quite so special as stuffed eggs. Herewith, three favorite versions from Arrowhead cooks.

PICNIC EGGS

To hard-cook eggs: Place eggs in cold water and bring to a simmer slowly, as this helps prevent bursting. Do *not* boil hard, as this toughens them. Simmer gently 15–20 minutes, turning several times to help center the yolks. Cool by plunging immediately into cold water—this helps eliminate the greenish ring which otherwise forms around the yolks.

POLLY'S DEVILED EGGS

Hard-cook desired number of eggs, peel and halve lengthwise. Remove yolks and combine with mayonnaise, Dijon and American-style mustard, salt and pepper to taste, working mixture with a fork until smooth. Mound in whites and finish with a sprinkle of paprika. For picnics, these pack easily if you fit two halves together again and wrap individually in foil.

CHARLOTTE'S STUFFED EGGS

8 hard-boiled eggs   *1 tsp. dry mustard*
1 tsp. vinegar   *Mayonnaise, salt and pepper to*
                             taste

Slice hard-boiled eggs in half lengthwise. Remove yolks into bowl and crumble with fork. Mix dry mustard and vinegar to form paste. Add to egg yolks with mayonnaise, salt and pepper to taste. Add 1 medium stalk celery finely chopped, or three slices crisp bacon, or 1/3 cup finely chopped ham. Stir together with fork and mound into egg whites.

SUPPER EGGS

For a super Sunday supper, place stuffed eggs in a baking dish. Melt 2 tbsp. butter, blend in 2 tbsp. flour. Add one 10 oz. can cream of shrimp soup and 1 can milk. Cook until thickened. Add 1/2 cup shredded sharp cheddar. Pour sauce over eggs. Bake at 350° 15—20 minutes.

CHOCOLATE DROP CAKES WITH MARSHMALLOW

2 squares baking chocolate   *1 cup sugar*
butter size of an egg   *1 cup flour*
2 eggs   *1 tsp. baking powder*
                1 tsp. vanilla

Melt together chocolate and butter. Add to eggs and beat vigorously. Add the sugar. Add gradually flour and baking powder sifted together and, lastly, vanilla. Bake moderate oven 350° after dropping by spoonfuls on cookie sheet.

Remove from tin while still warm, and then when cool put together with Marshmallow Fluff.

This may also be made with 4 tbsp. of carob powder instead of the chocolate, which gives the cakes a distinctive, delicate flavor.

CHEWY LOUIES

½ cup butter	1 tsp vanilla
1½ cups lt. brown sugar	2 eggs
1¼ cups flour	1 cup chopped nuts
¼ tsp. salt	1 cup (3½ oz.) coconut

Cream butter and 1/2 cup sugar. Add 1 cup flour, 1/4 tsp. salt, and mix well. Put into 3″ × 9″ × 2″ pan. Bake in moderate oven 10 minutes. Mix 1 cup brown sugar, 1/4 cup flour and remaining ingredients. Sprinkle evenly over the partially cooked mixture. Return to oven and bake 20 minutes more. Cut into squares while warm.

FALL

"Not yesterday I learned to know
The joy of bare November days
Before the coming of the snow . . ."
ROBERT FROST

SOME years it comes in late August; other years not until mid-September: that day when suddenly the soft haze of summer is gone, the clear air has a tinge of gold in it, and the goldenrod that seemed green yesterday blazes fully yellow in the new light. The sun is warm, but there's an unmistakable crispness to the air, too.

Frost comes late at Arrowhead, and summer crops linger on, protected by warm breezes from the nearby ocean. But the brisk wind down the Merrimack and the ever-earlier darkness tell us that summer is at an end.

Ever since the first inhabitants measured the season by its moons—the Harvest Moon, the Hunter's moon, the Moon Before Snow—fall in New England has been a season of both completion and preparation. The stubbled fields, the drying foliage burnished to reds, golds, and browns, the sounds of dropping nuts and the smells of winy fruits unmistakably mark the end of the growing cycle. But as nature "winds down," there is a sense of urgency about getting ready for winter. As the wild geese fly overhead, New England farmers secure the fruits of their year's labor.

In the old days, fall meant harvesting and processing the hard, dry flint corn which would supply meal for the next year's Indian cake, brown bread, and hasty pudding; drying slices of pumpkin and apples for use in pie and sauce; butchering livestock and salting and smoking the meat. It also meant pressing cider, some for immediate use sweet, the rest for the year's supply of hard cider. Uncle William's 1850s diaries contain many references in November to pressing cider for his family and for neighbors, and to selling it in town—in one case, enough to buy a new gundalow. And the house had to be readied for cold weather—the parlor stove brought in from storage and set up; salt hay mounded against the house foundations and secured with planks; firewood to be split and stacked in an orderly fashion, with kindling, hardwoods, and pines easily accessible for different purposes in the woodshed.

Today at Arrowhead, fall means picking the pumpkins and winter squashes; picking, cleaning, and polishing gourds; picking, curing, and bunching dried flowers and Indian corn; then shipping truckloads of them to the Boston market, and selling wagonloads more at local harvest festivals and fairs, and at the farm stand.

The main crops—and their uses—have changed somewhat in the twentieth century, but many of the fall activities at Arrowhead today remain the same as in earlier years. Tomatoes still abound in September. We hurry to get them in and use them before they freeze, and the kitchen is again

Fall, 1890's. LMP

filled with the pungent smells of the last batches of tomato soup, relish, and chili sauce. The purple Concord grapes ripen in the September sun behind the house, and juice and jelly simmer in big kettles. And outside there is much to be done to prepare for winter: fall tillage, mulching, the planting of winter cover crops, and a seemingly endless round of covering up, bringing in, battening down. The work gets increasingly chilly as the season progresses, and the warm smells emanating from the kitchen carry an especially satisfying message at this time of year.

It is no accident that the quintessential New England holiday, Thanksgiving, comes in the fall. For this is the peak season of abundance in this land with its short growing season. Corn continues into October; tomatoes, onions, sweet peppers, potatoes, lima and shell beans, pumpkins and squashes of all kinds come from the vegetable garden; pears and apples from the orchard; cranberries from the bog. These are all among the most traditional of New England foods, and they form the basis for many of the hearty and flavorful dishes we associate with fall. Heartier foods—soups and stews, hot breads, puddings and pies—and spicier flavors taste good at this time of year. Many of New England's favorite recipes combine this native farm provender of fall with spices and flavorings from the East and West Indies to create food to warm both body and soul.

Fall, 1980's. PCH

 No matter how hard we try, we can't keep ahead of the demand for fresh corn. (Right): Dick takes farm wagon to Newburyport Harvest Festival on restored Inn Street. CDC

TEA TIME

With fall came an increased burst of activity for church-related organizations: it was time to get ready for Christmas. The Home Circle met to sew items to sell at the Christmas Fair, and the Fair Committee met to plan the event. The Missionary Society met to plan the gathering of clothing and other items to pack and send to the needy in the Southern mountains. All of these activities, of course, required the serving of tea after an afternoon of productive work. And tea required the best bone china cups and saucers, coin silver spoons, and fine linens. Each hostess was famous for her own specialties. At Arrowhead these included tiny sandwiches of pineapple and cream cheese or chicken, or date-nut bread with cream cheese; pie-crust-and-jelly tartlets, jelly roll, Sarah's cornflake macaroons, chocolate Jumbles, and Aunt Hannah's brownies.

SARAH'S CORNFLAKE MACAROON

2 egg whites, beaten stiff
1 cup sugar, beaten into whites

2 cups cornflakes, crushed a little
½ cup shredded coconut (optional)

Stir all together and drop by spoonfulls onto a buttered cookie sheet. (Recipe should make approximately eighteen). Bake in a moderate oven 350° until well set up but not too dry—about 12 minutes. Remove from tin and cool on wire rack.

CHOCOLATE JUMBLES

½ cup butter
1 cup sugar
2 sq. baking chocolate, grated or chipped

2 eggs, beaten
2 tsp. baking powder
2 cups flour, and enough more to roll out

1 tbsp. milk

Melt together butter and chocolate; when cool add sugar and beaten eggs. Sift together flour and baking powder and add to mixture with milk. Roll out on floured board to ½ inch thickness. Cut out with doughnut cutter, sprinkle with granulated sugar and place on buttered cookie sheet. Bake about 10 minutes in a 350° oven until firm.

AUNT HANNAH'S BROWNIES

1 cup sugar
½ cup butter
2 sq. baking chocolate, melted

2 eggs, beaten
1 cup flour
½ tsp. baking powder

½ cup walnuts, chopped

Cream together butter and sugar. Add melted chocolate and well-beaten eggs. Add flour and baking powder sifted together and, lastly, walnuts.

Bake in small 8" tin 350° oven for 30–35 minutes. Cut in squares before removing from pan.

JELLY ROLL

Use recipe for Never-fail Sponge Cake and bake in a large, shallow pan about 7" × 11." As this will be a thin cake it will bake faster, so check your 350° oven in about 12 minutes. Cake should still "sing" slightly when ready to remove from oven.

Dampen a towel and sprinkle it with confectioner's sugar; lay it on a large cake rack. Turn the cake pan upside down onto the rack. As soon as cake is removed from pan, carefully start rolling it up with help from the damp towel. When you have formed it into a roll, let it cool with the cloth over it. Then unroll the cake, quickly spread it with the jelly of your choice, and roll up again. When fully cooled it can be easily cut into thin, delicious slices.

BRINGING IN GARDEN PLANTS

When the blue-jays begin to flash about the yard sounding their fall bugle note, summer has passed. It's time to think of bringing in some plants from the garden to cheer our winter kitchens. But we will not disturb many. Only a few species perform well indoors; the rest, having bloomed extravagantly all summer, wish only to settle into the leaves and vegetate.

Probably a few geraniums for the south windowsill come to mind first. Thrust the garden fork in deep beside them and pry up, lifting the plant carefully so we break no roots. Now brush away much of the soil from roots, and place plant in a 4" pot, being sure to arrange roots in all directions. Holding main stem in center of pot, fill in soil to about 1" from top, setting crown of geranium a bit below soil level. Press firmly to seat plant, and water. Now, being completely heartless, we trim back all branches so only 2 or 3 leaves are left on each—yes, buds and all.

We find a shady place for them to recuperate and put them back into the sunshine after a few days. We must remember to keep these plants watered. Plunging the pots into the soil will help hold the moisture. Now, when Jack Frost is expected they will come into our cozy kitchen.

Fibrous rooted begonias are also a good choice. Trim these back also, or they will soon be spindly. It takes a while for all cut-back plants to renew themselves, but the results are worth it. All these winter friends will enjoy cool temperatures, which shouldn't be too hard to provide these days.

The smaller herbs respond well to indoor life and provide taste excitement. Rosemary will bloom happily in a southwest pantry window, as will chives and the lovely little pelargonium crispum, the finger bowl green of former days, with its spicy tiny leaves. Lemon balm and thyme of different sorts do nicely. With a bit of coaxing, lavender, particularly French lavender, will tolerate us. Remember herbs love sunshine. The mints generally like a damper atmosphere than our houses provide. For sage to season our poultry stuffing we must dig through the snowbanks.

In general, house plants require little food or water during the waning cycle of the year. With the stronger sun of mid-January, however, we begin to feed them more, organic Electra being a good choice, and by February we give them more water as required. Check the herbs carefully, though, as they sometimes like more water than we realize.

So do not be too sad when Jack Frost comes. Japanese gardeners say that only when the flowers are gone does our true garden emerge. The form is clearly visible, as is the texture of brick and stone, the line of arch-

Tulips and herbs add a colorful touch to the winter kitchen

ing shrubbery, and the motion of winter birds—all to be enjoyed through the window. Inside we will be content, watching our few plants grow and bloom while we plan next year's garden flowers.

Rose hip jam is one of the great delicacies of life. And the expedition to pick the rose hips is one of the great joys of September. The luxuriant *rosa rugosa* bushes are still verdant in the bright early fall sunlight, the sun warm and the breeze from the sparkling ocean cool. A few late roses cling to the thorny branches as a reminder of summer, but the hips are big and fat and bright orange. They should be picked before they freeze for best flavor and quality, but, if you can't make the jam immediately, they will store perfectly well in a plastic bag in the freezer.

Preparing them for the jam is time consuming, but the product is well worth it. This recipe will jell nicely if you follow the timing exactly, and the brief cooking time also helps preserve the vitamin C content, which is very high. The jam is delicate and distinctive in flavor. This recipe came to us from a friend, Sumner Thompson, who swears by Martha's Vineyard rose hips, but we've always been partial to those from Maine.

UNCLE SAMMY'S ROSE HIP JAM

Cut ends off rose hips and scoop out seeds. Add 3/4 cup water to each cup of rose hips. Puree in blender. Transfer to cooking pot. Add 1/2 cup lemon juice (2 lemons) for each cup rose hips. Boil 2 minutes. Sieve to remove seeds. Return to pot, add 3 c. sugar and stir to dissolve. When boiling, use pectin according to directions on bottle, 1/4 bottle per cup rose hips. Pour into sterilized jars and seal.

GRAPE JUICE

4 qts. Concord grapes removed   *4 cups water*
   from stems   *sugar*

Rinse and thoroughly crush grapes. Place in large kettle with water. Bring to boiling point slowly and cook until soft, about 10–12 minutes. Strain. For every quart of juice, add two cups sugar. Boil together 10–15 minutes until juice is slightly syrupy. Seal at once in pint jars as described in strawberry preserves. Store in cool place. When served, dilute the juice one-half, or to your taste.

WHITE GRAPE JUICE

If you are blessed with a white grape vine, do bottle up some white grape juice also. For this, proceed as for Concord grapes, but for every quart of juice add 1 3/4 cups sugar and 1 tsp. fresh lemon juice. Over the years our old White Niagara vine has produced many jars of this refreshing Nectar of the Gods, much to Elizabeth's delight.

CONCORD GRAPE JELLY

Select grapes which are ripe enough to be sweet but are still firm and a little tart. Remove grapes from stems and thoroughly crush 3 1/2–4 lbs. fruit in a large kettle. Cook slowly until juice is extracted. Strain; measure juice and allow 2 cups of sugar for each pint of juice. Gently heat measured sugar in oven while juice boils 20 minutes. Add warm sugar to juice and boil hard for 3 minutes. Skim, and pour into sterilized jelly glasses. Seal with paraffin and let stand undisturbed until cool and stiff.

White Grape Jelly may be made in the same fashion, but remember to add 1 tbsp. fresh lemon juice. As this is always a little softer jelly, you might like to use Certo, in which case follow the rules for Concord jelly but still add the lemon.

Alternatively called "November jelly," this is an old recipe, and one we've never seen anywhere else. It makes an exceptionally clear, golden-red jelly with a flavor that is the essence of fall. A colorful jar of it makes a lovely Christmas gift.

PARADISE JELLY

6–8 apples	*1 qt. cranberries*
6–7 quinces	*sugar*

Wash and wipe apples, cut into pieces, including core. Barely cover and cook slowly until soft. Pour into jelly bag and drain thoroughly, squeezing very lightly. Repeat process with quinces, which require longer cooking time. Rinse cranberries and cover with water. Cook very slowly until soft. Strain juice through bag and add to other juices. Measure juice, allowing 2 cups sugar to every pint of juice. Boil juice 20 minutes. Then add the sugar and stir until dissolved. Boil 5 minutes, or until it jells. To test for jelling, spoon a few drops onto a cold plate, and when it thickens fairly soon it is ready to put up. Pour at once into hot sterilized jelly jars and cover with melted paraffin.

Take heed how you handle paraffin. Melt it slowly and never leave it until you have removed it from the heat, as it is highly flammable.

SQUASHES, PUMPKINS AND GOURDS

Squashes and pumpkins have provided important sustenance for generations of New Englanders. The Indians grew them long before the white man came. New England farmers continued to develop new varieties, and New England cooks find new uses for them. The Indians also grew gourds to use as containers and utensils, but since then, they have been grown only for ornamental purposes.

Few vegetables, in fact, have changed more over the centuries than pumpkins and squashes. When the Pilgrims landed, they found cushaw and neck pumpkins and skinny, crooknecked squashes. By the mid-

nineteenth century, these had been transformed into field and sugar pumpkins and many new varieties of squash in all shapes and sizes.

Throughout the nineteenth and early twentieth centuries, the New England market wanted large squashes: Golden Delicious, Blue Hubbard, Boston Marrow; the larger, the better! But after World War II, as people's styles of living changed, their squash preferences changed, too. Today's fall and winter market is for "consumer size squash," such as Butternut, Acorn, and Buttercup: use the whole fruit in one serving, with no leftovers. Herewith, some comments on several of our favorite varieties.

Butternut—good all-purpose, nice "nutty" flavor.

Acorn—cut in half and bake with honey, brown sugar, or maple syrup glaze.

Buttercup—nice quality, dry, good texture, extra good for late winter storage.

Spaghetti Squash—new and popular variety; we recommend using as squash, not as spaghetti.

Blue Hubbard—large size (30+lbs.), traditional favorite. Cut with ax or toss down cellar stairs to split.

Pumpkins, which kept many an Indian and Pilgrim alive through the winter, have become, ironically enough, "ornamentals" in recent years, prized for their bright color and pleasing shape—and because they symbolize the bounty of fall and the Indian and Yankee heritage. At Arrowhead, we grow our own strain of field pumpkins for "Jacks," developed over the years for good orange color, nice round shape, better keeping qualities, and good sound "handles" (better stem retention). We also grow, for an increasing demand, two varieties of New England sugar pie pumpkins: Spookie, 6–8 pounds, and Young's Beauty, 10–12 pounds. Both varieties have thicker flesh, a finer texture, and a higher sugar content than field pumpkins and are great for pies, "custard cups," stuffing and baking, or serving cold soups from—the latter two being new uses our customers are discovering.

This year we will be growing a variety of "neck pumpkin" of Pennsylvania-Amish origin. Shaped more like a crookneck squash, this pumpkin should be of excellent eating quality.

As ornamentals, both pumpkins and gourds are today a main fall crop at Arrowhead. As is true with other ornamentals, they have become more profitable than "edibles" in recent years. We have spent a great deal of effort in developing our own varieties of these two crops, selecting, drying and saving the best seeds each year. Pumpkins are simply picked and sold; gourds are dried, brushed, and treated with a vinyl preservative. Every fall finds chainstore buyers, as well as our own customers, eager for our "top of the line" product. Fortunately, gourds are prolific, and there are usually enough for everyone.

SQUASH PIE

1½ cups squash	2 eggs, beaten
2 cups milk, scalded	cinnamon
¾ cup sugar	nutmeg

Select squash with care. Golden Delicious are still good, but those charming little Buttercups are even better. Ripe squash have a dried stem; the interior should be firm, thick-meated, and highly colored.

Cut squash in pieces with heavy knife that has proved itself worthy. Place in large kettle and boil until soft, about 25 minutes. Scoop meat from shell and put through potato masher.

Line shallow 9″ pie plate with a crust, roll edges under, and flute a bit as for pie shell. Prick bottom and slide into 400° oven for about 5 minutes.

Meanwhile scald milk in large saucepan, add squash and sugar and stir in thoroughly. Remove from heat and add beaten eggs. Pour into pie plate, sprinkling top with cinnamon and nutmeg freshly ground. Bake at 375° until firm all over. Do not allow to boil.

It's nice to cook extra squash for use later as a vegetable or to make

SQUASH CUSTARDS

These are made with the same recipe as for pie, poured into custard cups and baked till firm in 375° oven. Anyone home will gladly help you eat them right from the oven.

BAKED ACORN SQUASH

acorn squash	*water*
brown sugar	*butter*

Select small-to-medium acorns, as the thinner meat cooks faster and sugar will not brown too much. Cut squash in half, remove seeds and stem, and level pointed end a bit. Pour small amount of water in glass pie plate and place halves in it. Put about 2 tbsp. water in each half and sprinkle 2 tbsp. brown sugar into each cup and on the edge around it. Bake at 350° about 1 hour or until squash is soft, basting syrup up onto edges occasionally. Dot with butter a few minutes before removing from oven. Serve in its shell.

Arrowhead cooks insist that shell beans, unlike most vegetables, need a long slow cooking to properly release their flavor. Although they are boiled, the beans don't want to taste boiled and half raw. Prepared well, they have somewhat the consistency of baked beans, but retain their own robust flavour, part of which escapes into the semi-thick juice surrounding them.

SHELL BEANS

Buy the brightest red shell beans you can find, shell them, and pop them into boiling water, not too deep. Now turn down the heat so they cook slowly. Keep your eye on them, adding more water as needed. A burned bean is a ruined bean. Simmer at least 45 minutes or until juice is thickened and rich. Salt to taste. Add butter. These beans will warm the cockles of your heart on a crisp fall day. Many insist that they're even better warmed over, so do make a large potful.

Lima beans, the large ones, used to be considered one of the fall's blessings, but they have all but disappeared from the market. They do take a little longer to mature than other varieties, but seed houses still list them in their catalogs. If you garden, you might curb your impatience and enjoy a real treat. Be sure to buy only Fordhook seed. There is no substitute.

FORDHOOK LIMA BEANS

Wait for them to fill out their pods and ripen, but not dry, on the vine. Then pick them; put water on to boil while you shell them. Cook slowly for about 30 minutes, renewing water as needed. Salt lightly and butter generously. They are hearty beans and make a good main supper dish.

DICK'S FALL VEGETABLES

Equal parts cubed carrots, yellow turnip and potatoes; a few small quartered onions. Steam (but do not oversteam). Serve with butter, salt, and fresh-ground pepper. 10–12 minutes.
Can use parsley or basil.

STUFFED PEPPERS

Rinse 8 medium-size peppers, remove stem with sharp paring knife, and scoop out all seeds. Cover peppers with boiling water and let set for five minutes. Place in buttered baking dish and make a stuffing of 2 cans of Spanish rice (15 oz. size) and one-half pound of hamburger.

Heat Spanish rice till it bubbles. Meanwhile, crumble hamburg into little chunks and brown quickly in hot frypan, turning meat once so it cooks evenly. Mix meat into rice and stuff into peppers. Bake in over 400° about

45 minutes or until slightly browned on top. Serve with butter pat on each.

Small size peppers are also good stuffed with poultry dressing as a vegetable complement to meat.

HEARTY MEAT DISHES FOR FALL

PAULA'S CONGLOMERATION

1 lb. hamburger	*1 lb. carrots*
10–12 medium potatoes	*2 or 3 yellow onions*
1 can kernel corn	*butter, salt and pepper, milk*

Peel potatoes and carrots. Dice carrots coarsely. Put potatoes, carrots, and sliced onion in kettle, cover with water and cook until potatoes are soft. Drain off water. While potatoes are cooking, brown hamburger in skillet. Pour off fat. Whip potatoes, carrots, and onions together with 1/4 lb. butter, milk, salt and pepper until fluffy. Fold in hamburger and serve good and hot.

CHARLOTTE'S FALL CHOP SUEY

1 lb. hamburger browned in little chunks in hot frying pan.
Drain off as much fat as possible, and set aside meat.
2 tbsp. butter
8–10 tomatoes, cut to size of walnuts
4 lg. stalks celery sliced thin
Melt butter in large fry pan. Add tomatoes and celery. Cover and let stew together 10 minutes. Then add:
2 lg. sweet peppers, diced coarsely
4 onions, chopped coarsely
1½ cups snapped yellow beans
Simmer all together, uncovered, till vegetables are cooked and fragrant. Add 1 sm. can Hunt's Tomato Sauce, 2 tbsp. catsup, browned meat, salt and pepper to taste.

Pour 1/2 c. water into the meat fry pan, heat and stir to get brown juices and pour into chop suey. Let set 10 minutes.

Serve steaming hot with oven-crisped pilot crackers.

Justin named this version of American Chop Suey several years ago, and it remains a family favorite.

MERRIMAC SUEY

1 lb. hamburger
1 clove garlic, mashed
1 lg. can crushed tomatoes in puree
dry red wine—"about that much"
½ tsp. basil
2 tsp. parsley
½ tsp. black peppery
8 oz. large elbow macaroni, cooked

1 lb. yellow onion
1 green pepper, diced
1 12 oz. can tomato paste
2 bay leaves
1 tsp. crushed oregano leaves
1 tsp. salt
1 can sliced mushrooms
parmesan cheese

Slice onions. Brown hamburger with onion and garlic in large kettle. Add green pepper and cook 1 minute. Add tomatoes, seasonings, mushrooms, and red wine. Let cook slowly, stirring frequently about 1/2 hour. Add cooked macaroni (more if needed) and a generous amount of parmesan cheese. Serve good and hot, for several days.

HERBED MARINADE FOR MEAT

1 cup red wine
1 cup olive oil
¼ tsp. garlic powder

¼ tsp. each rosemary, thyme,
marjoram
1 tbsp. parsley

Combine ingredients and marinate meat for at least two hours before cooking. White wine can be used for poultry.

HERBED PORK CHOPS

Brown 4–8 chops in butter in heavy skillet. Remove to plate. Add to skillet 1 can cream of mushroom soup, 1/2 can water, 1/2 can red wine. Add herbs to taste. We especially like 1/4 tsp. rosemary, 1/4 tsp. marjoram, and a pinch of thyme. Add chops to sauce, cover, and simmer 1/2 hour. Serve with mashed potatoes.

ROAST PORK WITH ONIONS AND POTATOES

Ask your butcher to partially saw the bones down the length of a rib roast of pork so you can open the roast to form a little tent in a shallow open roast pan. Pour about 2 cups water around it, and place in a 425° oven for

1/2 hour. Then lower thermostat to 350° and continue roasting for 1 hour.

Slice 4 or 5 onions and 5 or 6 potatoes, placing both under the meat and around it. Raise temperature to 400° and continue roasting for 45 minutes to 1 hour, renewing water as necessary, basting meat and turning potatoes once. Let juice simmer down to brown vegetables and meat.

BAKED STUFFED PORK CHOPS

Select thin rib chops. Place them upright with bone down in buttered Pyrex loaf dish. Between each chop and at both ends of the dish, press small handfuls of bread and cracker crumb stuffing, as follows:

2 slices bread, crumbled	*2 fresh sage leaves, cut fine, or*
5 crackers, crumbled	*½ tsp. dried sage*
1 small onion, chopped fine	*bit of salt and pepper*
1 tsp. Bell's seasoning	*1 tbsp. butter*

Gradually blend all together, adding about 1/4 cup hot water slowly and mixing with fork. Bake in hot oven 425° about 1 hour.

FROM THE APPLE ORCHARD

Apple pie has long been a staple in the New England diet. In the eighteenth century, students at Yale received one apple pie each evening as a major part of their board, and this was considered by all to be quite normal. The traditional New England pie was hearty and nourishing (no chocolate chiffon or Bavarian cream, these!) and not overly sweet: apple, squash, pumpkin, mincemeat, and cranberry were fall and winter favorites, made with ingredients which lasted through the winter. New Englanders have always had a penchant for pie for breakfast, a custom which Ralph Waldo Emerson staunchly defended. "What is pie for?" he demanded. And, when the pie rack is nicely full, nothing is as easy, or as satisfying, to start the day as a good slab of pie with a steaming cup of coffee.

> *"The pie should be eaten while it is yet florescent, white or creamy yellow, with the merest drip of candied juice along the edges, (as if the flavor were so good to itself that its own lips watered!) of a mild and modest warmth, the sugar suggesting jelly, yet not jellied, the morsels of apple neither dissolved nor yet in original substance, but hanging as it were in a trance between the spirit and the flesh of applehood . . . then, o blessed man, favored by all the divinities! eat, and give thanks, and go forth, 'in apple-pie order!'"*
>
> HENRY WARD BEECHER

NORTHERN SPY APPLE PIE

Northern Spy apples are ready for use in late fall, and last right through February. They are noted for their fine flavor and good keeping qualities.

6–7 Northern Spies, medium size	*½ tsp. fresh grated nutmeg*
1½ cups sugar (¼ cup more if apples are not too ripe, or to taste)	*½ tsp. cinnamon*
	⅓ cup water
2 tbsp. flour	

Roll out pastry for a two-crust pie. Line 9″ pie plate. Peel and core apples. Slice into pie plate, making thin slices. Distribute sugar evenly over apples, add water, and sift flour over top. Sprinkle with cinnamon and nutmeg. Arrange top crust, moisten edges, and seal. Bake as previously described. Serve with sharp cheddar cheese.

APPLESAUCE

Each variety of apple has its own special flavor as well as season, and cooks prefer certain ones for certain recipes. Aunt Ida said, "Gravensteins for jelly, and Macintosh for fall sauce."

So let's cut up ten or twelve Macs with their red skins left on; remove core and pop them into a large saucepan. Barely cover them with water and cook slowly until they are soft. Then put through a sieve to remove the skins, which have served their purpose of tinting the sauce with ruddy color.

Save out a cup of this sauce, before sweetening it, to make Applesauce Cake. Add a cup of sugar to remaining sauce while still hot, and a speck of cinnamon if you will, or leave the flavor pure apple. Cool it, and it's good with many things, among them thin white pork chops and hashed brown potatoes.

Later in the fall, try Cortlands and Northern Spies, both delicious in sauce.

APPLESAUCE CAKE

1 cup unsweetened applesauce	*2 cups flour*
1 tsp. soda dissolved in little water	*½ tsp. allspice*
	½ tsp. cinnamon
1 cup sugar	*¼ tsp. nutmeg*
2 tbsp. butter melted	*pinch salt*
½ cup seedless raisins	

Beat dissolved soda into applesauce with spoon. Add sugar and melted butter and stir all together. Sift dry ingredients together and add to applesauce mixture. Stir in lightly floured raisins. Turn into buttered loaf pan and bake in 350° oven about 1 hour and 15 minutes.

This is a tasty loaf, quick to mix, and it keeps well, if given the chance.

In baked apples, as in apple sauce, your choice of variety is very important. Macintosh are good baked in mid-fall, followed by spicy Cortlands in late fall, or tangy Northern Spies. Although Golden Delicious are not commonly considered cooking apples, try some baked on a chilly evening. You'll enjoy their delicate, distinctive flavor.

BAKED APPLES

Wash and core apples of uniform size. Fit into a round glass pie plate and fill each apple center with about 2 tsps. sugar. Pour 1/2 cup water around apples. Bake at 350° approximately one hour, basting once or twice with the juice. When apples are soft and puffy in center and juice is syrupy, it's time to remove from oven. Serve warm with pouring cream for a scrumptious supper dessert.

BAKED APPLE TAPIOCA

3 cups water	*1 cup light brown sugar, firmly*
2 tbsp. lemon juice	*packed*
3 tart apples, pared and sliced	*½ tsp. salt*
½ cup minute tapioca	*½ tsp. fresh nutmeg*

3 tbsp. melted butter

Combine water and lemon juice and pour over apples in greased baking dish. Cover dish and bake in a 375° oven for 15 minutes or until apples are partially cooked. Mix together minute tapioca, 3/4 cup brown sugar, salt, and nutmeg. Sprinkle mixture over apples, stirring in well. Add melted butter. Bake 10 minutes. Then stir well, sprinkle remaining 1/4 cup brown sugar over apple mixture and bake 5 minutes longer.

Serve either hot or cold with light cream or half and half. Makes 6 servings.

BIRD'S NESTS

These little desserts are very appealing to the eye and the taste.

Core and pare 8 apples. Make ready round pieces of sponge cake, one for each apple, an inch thick and same diameter as apple.

Make syrup of 1 cup sugar and 1 cup water boiled 5 minutes. Place apples in it and cook them very slowly, turning each over once, until they are tender.

Dust the rounds of cake with sugar and place them in round pie plate in moderate 350° oven until sugar melts and runs over cake. Then carefully place an apple on each cake.

Now add half a glassful of currant, quince, or strawberry jelly to the syrup and cook until smooth and thick. Pour over apple cakes and serve with whipped cream.

ELIZABETH'S APPLE PUDDING

Butter well a qt.-sized bread pan (glass). In bottom arrange 2–3 large Cortland apples, pared, sliced, and cut into size of a half walnut shell. Sprinkle nutmeg generously over apples.

Combine 2/3 to 3/4 cup sugar with 2/3 to 3/4 cup flour and dash salt. Mix well. Into this dry mixture cut about 2/3 of 1 stick of butter—butter should be reasonably soft. Distribute evenly over apples.

Bake 20–25 min. in 400° oven or until slightly brown and apples appear soft. Serve hot or warm with plain cream, whipped cream, or vanilla ice cream. (If apples are used early in season, cut in smaller pieces—but as season progresses and Cortlands become more mellow, they can be cut larger. Cortlands are the only apple I use. If apples are quite hard and green early in season, you may sprinkle 1/4 cup water over apples before the topping.)

QUICK APPLE PUDDING

1 egg	*2 tbsp. flour*
1 cup sugar	*1¼ tsp. baking powder*
1 tsp. vanilla	*dash salt*
walnut meats	

Beat egg slightly. Add sugar and vanilla. Mix together flour, baking powder, and dash of salt. Peel and chop into rather small pieces 2 large apples (preferably Cortland, though others are o.k. for this recipe). Put these chopped apples in a greased glass pie plate. Then sprinkle broken walnut meats over apples. Mix the egg mixture and dry ingredients well and pour evenly over the apples. Bake about 350° for about 30 min.

Serve hot with cream, whipped cream, or vanilla ice cream. Makes 4 decent-sized servings. This pudding is quick to make and can be baking while you are assembling the main part of the meal.

CRANBERRIES

No other fruit is so completely associated with Massachusetts as the cranberry. Originally called "craneberries"—whether because the colonists saw the cranes eating them, or because their flowers resembled a crane's head and neck—they soon took an important place on Massachusetts tables. John Josselyn noted in 1663 that, "The Indians and English use them much, boyling them with Sugar for Sauce to eat with their Meat, and it is a delicate Sauce." Yankee ships soon carried them as a preventative for scurvy.

All too many people today know this bouncy little berry only through its very distant relative, canned cranberry sauce. More's the pity, for the real thing is not hard to make. And in addition to sauce, as the following recipes show, there are many other delightful ways to use cranberries.

You're fortunate if you have access to fresh water cranberries. They are a little smaller and rounder than their salt water cousins and are noted for very fine flavor. Alas, they are scarce.

Perfect cranberry jelly is not hard to make if you follow the directions exactly. If you don't, it will not jell. The timing is crucial; use the second hand of your watch rather than the range timer!

ELIZABETH'S CRANBERRY JELLY

Pick over and wash 4 cups cranberries. Put into saucepan with 2 cups boiling water; return to boil, cover and simmer 20 minutes. Put through potato masher, return to saucepan, bring to boil, and cook uncovered exactly 3 minutes over moderate heat—you want a moderate but not a racing boil. Add 2 cups of sugar, stir to completely dissolve, and boil fairly rapidly 2 minutes. Pour immediately into warmed glass dish. Once you have poured it into the dish it must remain absolutely undisturbed until it has cooled thoroughly and jelled; otherwise it will not jell. Allow at least four hours; preferably overnight. Or have 2–3 hot sterilized jars ready to fill. Pour cranberry jelly into these and seal *after* it has jelled. They will keep several months in your refrigerator.

WHOLE BERRY CRANBERRY SAUCE

Cook 4 cups cranberries and 2 cups water in covered saucepan over moderate heat until most of the berries have burst. Add 2 cups sugar, stir to dissolve, and cook 2 to 3 minutes longer. Remove from heat and chill. Excellent with pork as well as with the traditional poultry.

CRANBERRY PIE

A proper New England cranberry pie always comes to table dressed in diamond lattice coverlet. Line a 9″ shallow pie plate with pastry and roll out a sheet to cut into 3/8″ to 1/2″ strips for top.

Wash and measure 3 cups cranberries into saucepan. Barely cover with water and stir in 1/2 cup sugar. Cover saucepan and bring to a boil. You will hear cranberries "pop, pop"; when popping lessens, remove from heat, stir in 1 1/2 cups sugar and pour cranberries into pie plate. Sift about 1 tbsp. flour over top.

Moisten rim of crust with cold water and place pastry strips across pie, laying second course diagonally to form diamonds. Roll edge a bit and seal strips. Bake at 450° 10 minutes, lower heat to 375° and continue baking till filling begins to bubble.

This makes a nice, firm conserve. A good accompaniment for chicken, turkey, ham, etc.

ELIZABETH'S CRANBERRY CONSERVE

1 orange	*1 cup currants*
1 pt. water	*2½ cups sugar*
1 qt. cranberries	*1½ oz. walnut meats*
½ cup citron	

Wash and halve the orange, slicing very thinly into small pieces, using rind and juice. Seedless California orange preferred. Add the pint of water; boil slowly for 10 min. Cook cranberries with 1 cup water until very soft. Put through potato masher and add to cooked orange.

Add currants to orange and cranberry mixture with the sugar. Cook until mixture heaps upon spoon. Simmer slowly rather than boil. It may take 20–35 min. or longer. Stir in broken walnut meats and citron, diced. Mix well and pour into sterile glasses. Will keep in refrigerator for a long time. May be covered with paraffin or screw-on lid.

CHICKEN RUBY

2½–3 lb. cut up broiler   1 tsp. orange peel
½ cup flour   ¼ tsp. cinnamon
1 tsp. salt   1½ cups cranberries
4 tbsp. butter   ¼ cup chopped onion
¾ cup sugar   ¾ cup orange juice
    (1 can whole cranberry sauce can be used, but omit sugar)

Coat chicken with flour, salt, and brown in butter. Combine remaining ingredients, add to chicken and bring to boil. Cover and simmer 35–40 minutes.

CRANBERRY BATTER PUDDING

½ cup sugar   1⅔ cups sifted flour
1 tbsp. butter melted   2 tbsp. baking powder
½ cup milk   pinch salt
1 egg separated   1 cup cranberries

Beat egg yolk and blend with milk, sugar and melted butter. Sift flour, baking powder, and salt into a bowl and make a well. Pour in liquid, stir until smooth, then fold in stiffly beaten egg white. Stir in cranberries which have been cut in half and lightly floured.

Bake in greased 8″ cake tin or ironstone pudding dish in preheated oven 375° for 10 minutes. Lower heat to 350° and bake 30–35 minutes longer.

Serve with soft sauce flavored with lemon juice or vanilla.

SOFT SAUCE

2 cups sugar   3 cups water
2 rounded tbsp. flour   1 tbsp. butter
            1 tsp. vanilla

Work the flour and sugar together in a large saucepan; add water, stirring as mixture comes to a boil. Add butter; simmer 20 minutes; add vanilla, or flavor with fresh lemon juice, and serve hot.

CRANBERRY BREAKFAST CUPCAKES

½ cup butter	*2 tsp. baking powder*
2 cups flour	*½ tsp. salt*
1½ cups sugar	*2 cups whole cranberries*
½ cup milk	*2 eggs*

sugar for topping

Cream butter and sugar until fluffy. Add eggs one at a time. Sift dry ingredients together and add alternately with milk. Chop berries coarsely and mix into batter. Bake in cupcake papers, pile mixture high in each cup. Grease top of muffin pan. Sprinkle each cupcake with sugar. Bake at 375° for 25–30 minutes. Cool in tins before removing.

FORCING BULBS

Against the day when there will be no nuts on the trees, the busy squirrel in October stores away little caches of future sustenance, hiding them under the leaves to retrieve in bleaker days. So may we store away flower treasures for less cheery days when our courage needs renewing. October is the time to pot up some bulbs for winter.

Daffodils, Dutch hyacinths, little blue grape hyacinths, and crocus are among the easier flowers to work with. Tulips require a little expertise, as do iris. Unless you have a part of your cellar or attic which is cool, yet will not freeze, and a really sunny window, you should not attempt to force spring bulbs. "Forcing" means bringing flowers into bloom before their natural season.

Spring bulbs should be planted 3–5 of a kind together in a shallow bulb pan, clay or plastic. Ordinary garden soil may be used, placing a handful or two of soil in bottom of pan, then setting bulbs (right side up, please) on this and completely covering with soil. Level of soil should be about 1 1/2" from rim of pot. Label and thoroughly water each pan before storing in cool dark place—temperature about 45° is good. They will stay there till January or February so check for water occasionally, keeping them moist but not really wet. Do *not* let them completely dry out as they are not sleeping but are making roots for their later period of growth. When they are finally brought into your warm rooms—60° is fine—give them all the sunshine available to make sturdy plants.

There are many lovely daffs on the market, but it's wise to check with a knowledgeable garden center as to the best forcing varieties. Be sure to buy good grade bulbs, preferably double nose No. 1 and pre-cooled stock for earliest forcing. These can be brought into room temperature in late December if they show good root growth then; otherwise wait until early January. For continuing bloom bring in a pot every two weeks. You might like to give them one feeding of Electra during growth period, but most of their growing energy is already stored in the bulb.

Dutch hyacinths are beauties of pink, blue, and white, and will scent your entire house, but be patient and wait till February to start them up. This also applies to the tiny grape hyacinths. Both these species may be brought in at 2 week intervals. Don't permit any of the bulbs to linger in storage too late towards spring as they will mature very tall and spindly.

Lastly we will plant the daintiest member of all spring's *avant garde*, the paper white narcissus. These may be potted in soil like the others, but they are far more charming grown in a pottery bowl of pebbles and water. Place them in cool darkness the first week in November; bring them into warmth and light the first of December and enjoy them with the coming of the New Year. The blossoms eventually dry to a color and consistency of Japanese rice paper, but they open airy and white like little ballet dancers heralding the pageant of lengthening days and returning spring flowers.

THANKSGIVING AT ARROWHEAD

The main focus of Thanksgiving at Arrowhead, as everywhere, has always been The Dinner. Preparation for it began well ahead of time, with the baking of little citron cupcakes which were then placed in a crock to "ripen." For several days ahead, also, pies were baked, so that there would be an ample variety, while the oven would be free the night before and the day itself for the pudding and the turkey. The dining room, normally closed off at this time of year and conveniently cold, soon boasted a fine array of mince, squash, apple and fine-latticed cranberry pies on pie racks. The day before the great event, the kitchen bustled with activity as hard sauce, fudge, cranberry sauce, pudding, turkey and vegetables were assembled.

The flurry continued through the next morning as the dining room table was set, relatives arrived and last-minute dishes were readied. In fact, the kitchen barely quieted down during the meal itself, as constant new courses were hurried in. And after the last talk had died down, and the last possible bits of penuche and nuts had been consumed, it hummed again, if more quietly, as the women of the family combined visiting with cleaning up.

Back at the turn of the century, it seems, the ponds could generally be counted on to freeze over by Thanksgiving, and the day often ended with a skating party. More recently, and perhaps then for the older folks, a walk in the chilly gathering dusk has helped revive overstuffed spirits. If we are ambitious it could be an expedition to explore "Devil's Den" or look at a passing flock of Canada geese at the Plum Island wildlife refuge; if less so, a quiet stroll down the old lane to the site of the ferry will suffice.

THANKSGIVING MENU

Cranberry juice or Aunt Lizzie's Ambrosia
Roast turkey or pair of chickens w/cracker stuffing
Mashed potatotes w/gravy
Boiled Prince Edward Island turnips
Squash
Creamed onions
Peas
Cranberry sauce
Celery
Fruit bread
Parker House rolls
Thanksgiving pudding w/hard and soft sauce
Cranberry, mince or squash pie
Penuche, fudge
Mixed nuts
Coffee

AUNT LIZZIE'S AMBROSIA

3 apples chopped fine
3 bananas sliced
3 oranges chopped fine
½ lb. figs chopped fine
Juice of 2 lemons

1 sm. can (#10) pineapple slices
chopped fine
1 cup nutmeats chopped
1 cup shredded coconut
2½ cups sugar

Mix lemon juice and pineapple juice with sugar. Put fruit into dish in layers, adding a little of the sugar and juice to each layer. This makes enough for 10–12 fruit cups. Chill several hours, and serve in little stemware sherbets or in small punch cups.

ROAST TURKEY

Place turkey in covered roast pan with 2 quarts of water. Close vent in cover, place on top of stove over high heat and steam bird for 20 minutes. Remove cover, drain any water from turkey cavity and spoon in the stuffing, packing lightly so it will be fluffy when cooked.

Cook bird in covered roaster for 30 minutes, oven 425°. Turn heat back to 350°, open vent and roast for 1 hour. Remove cover, baste and continue cooking for 1/2 hour uncovered to brown and crisp the skin. Let the juice simmer down and brown. Check again and baste. Place a slice of bread over each wing and over drumsticks to prevent them from becoming dry and too brown. If necessary add water, but be sure to let juice brown well in order to make good gravy. Lower heat to 325° and roast for another 30–40 minutes. Check again, testing with cooking fork to see if meat is done as you like it. Put cover back on roaster for last 15 minutes.

Arrowhead prefers fresh turkey, 15–16 lbs., but frozen ones respond well to this same method. Consult time and temperature charts for different weights and vary accordingly.

When done to perfection, golden outside and tender and moist within, the flavor of this bird is pure turkey with a delicate hint of the simple stuffing. Place it with pride on your best platter and bear it to the table for the master of the house to carve.

TURKEY STUFFING

10–12 common crackers, crumbled *4–5 sage leaves fresh, or 1 tsp.*
4 slices bread, crumbled   *ground dried sage*
2 tsp. Bell's seasoning   *2 medium stalks celery, chopped*
½ tsp. salt   *1 medium onion, chopped*
freshly ground pepper to taste   *2 tbsp. butter*
½ cup hot water (approximately)

Mix cracker and bread crumbs together with celery and onion and seasonings. Add hot water and bits of butter a little at a time, mixing in with fork until stuffing seems consistency you like best. Now stuff the bird.

ROAST CHICKEN

Plan for a delicious roast chicken dinner and make one-half amount of stuffing for turkey and set aside.

Place a 4–5 lb. chicken in a covered roast pan with one quart water. Close vent in cover, place on top of stove over high heat and steam for

15 minutes. Drain any water from cavity and stuff chicken. Then place in preheated oven 375° and cook for 1/2 hour. Turn back heat to 350° and use time and temperature chart to determine roasting time according to weight of bird. Let juice in roaster simmer down and brown; watch it and add a little water if necessary. 1/2 hour before chicken should be done, remove cover and let bird roast and brown a bit, and juice brown onto bottom of pan. Your chicken skin will be a bit crisp and meat will be tender and juicy.

Bear your golden brown bird to the table and enjoy.

After dinner, add a quart or so of water to roast pan. Place over gentle heat, stirring while it simmers and gathers flavor. Add 2 tbsp. flour smoothed into a little cold water and cook in pan 2 or 3 minutes. Stock should not be too thick. Set aside for use in chicken soup.

CHICKEN SOUP

1 stalk celery chopped, throw in leafy end to be retrieved later
1 med. onion chopped coarsely
1 qt. water. Simmer onion and celery 10 minutes in water and add
4–5 potatoes sliced and cooked 10 minutes

Combine vegetables with soup stock from pan and add a generous amount of chicken meat left from roast. If there is a whole drumstick left or the neck, throw that in also as bones add sweet flavor. Cook together over low heat for 10 minutes. Remove from heat and add 2 or 3 rosemary needles and serve.

CREAMED ONIONS

tiny white onions (5–6 per person)   *heavy cream*
butter   *paprika*

Choose onions of a fairly uniform size. Blanch in boiling water for one minute; cool and peel. In a heavy flameproof serving dish, saute very slowly in generous amount of butter, turning frequently to avoid burning, until lightly browned and tender. Add enough heavy cream to form a generous sauce. Continue to cook slowly until sauce is thickened and golden brown. Just before serving, dust with paprika.

This recipe requires patience in the cooking and is outrageously rich, but so delicious that it's well worth both the trouble and the calories!

This recipe makes a firm, moist, sweet bread which keeps extremely well. It also travels well and is ideal for mailing as a Christmas gift. It is good plain or buttered, and also makes a delicious and different toast.

FRUIT BREAD

2 cups flour	*1 cup sugar*
1 tsp. baking powder	*½ cup milk*
½ cup shortening	*2 eggs*
1 cup candied fruit peel	*1 tsp. vanilla*

Sift dry ingredients. Work in shortening. Add fruit. Beat sugar, milk, eggs, and vanilla together; add to flour mixture and mix well. Bake in well-greased pyrex pan—360° first 1/2 hour, 325° second 1/2 hour. Bread is done when straw inserted in center comes out clean. Cool before slicing.

THANKSGIVING (AND CHRISTMAS) CRACKER PUDDING

Thanksgiving pudding is one of the central traditions of Arrowhead, and we have never come across it anywhere else. The closest recipe we've seen is in the original 1896 Fanny Farmer cookbook, but it is not identical.

It might be thought of as a derivation, more economical and suited to simpler Yankee tastes, of the old English Christmas plum pudding. But New Englanders celebrated Thanksgiving for nearly two centuries before the Puritan prejudice against the "pagan" celebration of Christmas died out, and, as the last stanza of Lydia Maria Child's 1847 poem, "Thanksgiving Day" shows, "pudding" was an established part of the menu in her childhood in the early 1800's:

> *"Over the river and through the wood—*
> *Now grandmother's cap I spy!*
> *Hurrah for the fun!*
> *Is the pudding done?*
> *Hurrah for the pumpkin pie!"*

When Arrowhead began to celebrate Christmas, sometime in the nineteenth century, "The Pudding," with only its name changed, moved onto the Christmas menu, too. And in one memorable year in the twentieth century, Polly and Glen somehow convinced Nana Moulton to make a New Year's Pudding as well.

The pudding itself seems to be an acquired taste, requiring several holiday seasons' exposure; but most people accept it readily as a vehicle for the hard sauce and we've noticed that, as the years go by, the ratio of pudding to sauce gradually increases. Dick, however, has never learned to like it, and prefers his hard sauce on anything else—he has even been known to have it on squash pie!

This recipe makes a large pudding. All the parts keep well in the refrigerator, and the pudding is even better warmed over.

24 common crackers	1 cup sugar
3 qts. milk	4 eggs. beaten
1 cup raisins	½ tsp. salt
2 tbsp. butter	

Crush common crackers. Combine with milk, raisins, sugar, eggs, and salt in buttered 4-qt. ironstone or glass baking dish. Dot with butter. Place in preheated 350° oven. Reduce to 325° after 1 hour. Bake for a total of about 3 hours stirring often, until it turns a golden brown. Do not stir during last 45 minutes, to allow crust to form. Serve warm, with hard and soft sauces.

HARD SAUCE

¼ lb. butter	2 tbsp. milk
1 lb. confectioner's sugar	1 tsp. vanilla

Cream butter; gradually add confectioner's sugar and milk. Add vanilla. Spoon into serving dish and chill until hard.

Mortar and pestle used to crush the crackers for holiday puddings

SOFT SAUCE

2 cups sugar   2 cups water
4 tbsp. flour   1 tbsp. butter
1 tsp. vanilla

Mix flour thoroughly with sugar; add water and butter. Bring to a boil, reduce heat and simmer 10 minutes, stirring occasionally. Remove from heat and add vanilla.

Thanksgiving and Christmas wouldn't be the same without sparkling cut-glass dishes full of these old favorites.

FUDGE

2 cups white sugar   ¾ cup milk
3 tbsp. cocoa   2 tbsp. butter
½ cup chopped walnuts (optional)

Mix cocoa into sugar in large saucepan. Add milk, stir to moisten sugar. Gradually bring to a boil and boil until a drop forms a soft ball in a glass of cold water—about 15–20 minutes. Remove from heat, add butter, and beat at high speed with electric mixer until mixture begins to thicken. If desired, add walnuts at end of beating. Pour into greased 9" pan (you will need to work very quickly at this stage, as fudge often "sets up" fast). When cool, cut into small squares.

PENUCHE

1 cup brown sugar (dark or   ⅔ cup milk
  light to taste)   2 tbsp. butter
1 cup white sugar   1 tsp. vanilla
½ cup walnuts

Proceed as for fudge, adding vanilla with butter.

Mincemeat was a staple for hearty winter pies. Every cook had her favorite recipe for it, and they varied greatly. Some included green tomatoes rather than apples; some had vinegar and some did not; some had meat and some did not. This recipe was Aunt Ida's favorite, and we like it, too. We assume the "plain" in the title refers to the fact that this recipe included no liquor. It is also good with a tablespoon or two of brandy added at the last minute.

PLAIN MINCE PIE

1 cup finely chopped meat
⅓ cup suet, chopped
2 cups chopped apples
½ cup seedless raisins
½ cup currants
½ cup chopped citron

1¼ cups brown sugar
1 cup sweet cider
juice of 1 lemon
1 tsp. each salt, allspice, cinnamon
½ tsp. clove, ground

2–3 dessert spoons jelly

Mix all together and simmer slowly for about an hour, stirring occasionally, adding a bit of water if necessary. Store in refrigerator for a few days. Then make your pie crust and you will have a couple of nice mince pies. This recipe can also be made in multiple quantities and sealed in sterilized jars for later use, or stored for several weeks in the refrigerator.

A full pie rack

Winter on Ferry Road: vehicles have changed since from the 1890s (above, LMP) and the 1940s (CMC), but snow still piles up.

"As the days begin to lengthen,
The cold begins to strengthen."
OLD FARM ADAGE

WINTER

WINTER, when the ground is frozen tight and blanketed with snow, is the "slack time" on the farm. With house and farm secured against the weather's onslaught, it is a time to sit near the kitchen fire with the coffeepot on, taking stock of last year, planning for next year, visiting with the family.

Which is not to say that there is no work to be done. Even on the stormiest days, livestock must be fed and provided with water to drink—which in bitter cold weather often means carrying hot water to the barn, and cows must be milked. This is the season when the New England farmer most appreciates the connected house and barn designed by his forebears to make these daily chores easier in inclement weather.

Traditionally, this has been a time for mending harnesses and repairing machinery and for making improvements around the house, as well as for turning to money-making trades such as shoemaking, silversmithing, and furniture making. Milling and selling grain, selling hay, cider, and other stored crops also brought in cash through the winter and wrapped up the year.

Winter has traditionally been woodcutting time on the New England farm, too. With the underbrush and leaves out of the way, and snow on the ground to make dragging the sledge easy, the farmer could conveniently select, cut, and bring out lumber for sale or building projects, firewood, and fenceposts for spring fence repair. This is still a winter activity at Arrowhead. And in early winter, the ritual tramp over the place to find the perfect cedar for the Christmas tree and gather pine boughs, cones, and princess pine for our house has escalated into a commercial operation to supply greens for other people's houses.

After Christmas, with the farm stand closed, comes the time, much as in the past, for wrapping up last year's operation, selling hay and other stored crops, catching up on the complicated bookkeeping which must be done before tax time, and to plan for next year—reading up on new varieties, planning acreage, ordering seed. It is also a time for attending agricultural seminars to find out about new techniques and discuss common problems with other farmers. In recent years, one course of seminars has become a requirement for renewal of a license to handle pesticides. And, as in the past, it is time to undertake "put off" projects around the house, repair machinery and clean up storage areas.

Christmas, of course, brought—and still brings—family reunions. Even before New Englanders began to celebrate what the Puritans considered a pagan holiday, they took advantage of the frozen rivers and snow-smoothed roads, piling into sleighs to visit distant friends and relatives in the winter. Travel by sleigh was fun, albeit cold, and farm folks often organized expeditions just for the ride; they also gathered parties for skating on the frozen rivers and ponds.

Appetites whetted by hard work and play in the cold, and many visitors around, called for plenty of hot, hearty food. While some raw materials, especially in the fruit and vegetable department, were limited, the larder so diligently filled during the summer and fall provided plenty of basics, and New England cooks used what they had inventively to create an abundance of tasty, warming food. "Indian" meal ground from the flint corn, and flour milled from other grains became the basis for hearty puddings, pies, cakes, and cookies flavored with molasses, spices, raisins, and chocolate from the overseas trade. Dried or stored fruits, especially apples, pumpkins, or cranberries, and preserved fruits of many kinds went into puddings, pies, and sauces, too. Beef from the late fall or early winter butchering, sometimes preserved by freezing in the cold weather, provided roasts for Christmas and Sunday dinners, and the tougher cuts became meat puddings and stews.

With the men folk around the house, and the inevitable days when everyone was snowbound, winter was always a time for special family cooking treats at Arrowhead, and it still is, for there are still many winter storms when it takes the city plows a long time to reach the end of Ferry Road. These are days to gather around the kitchen table drinking coffee or cocoa and helping to make doughnuts or cut out gingersnaps in fancy shapes, or watch the apple fritters fry, and ignore the wind and snow battering at the corners of the house.

SNOWY DAY FAVORITES

When snow sweeps across the fields and swirls in drifts in the yard, the kitchen seems especially warm and inviting. Children home from school and menfolk unable to work outdoors prowl hopefully around, looking for something to do, or eat, or both. Sooner or later, someone will get

A private tea party on a winter day, c.1919. LMP

out the rolling pin and pie-board for gingersnaps, or wheedle the cook into making some other special treat.

Making gingersnaps with a wonderful variety of cookie cutters—and freehand shapes encouraged—has always been a favorite snowy-day activity of Arrowhead children.

ROLLED GINGERSNAPS

1 cup molasses	*3 cups flour*
½ cup shortening	*1 tsp. soda*
	½ tsp. ginger

Simmer molasses and shortening together until shortening is thoroughly melted and mixed with molasses. Turn hot mixture into dry ingredients, which have been sifted together. Mix thoroughly. Cool. Roll very thin on floured board, and cut into shapes. Bake on well-greased pan at 375° for 10–12 minutes. Cool slightly before removing from pan.

HOT BREAKFASTS FOR COLD MORNINGS

OATMEAL

Oats were grown in New England from the earliest days of settlement, but it was not until the eighteenth century that they were considered fit for human consumption. From that time on, oatmeal became a favorite breakfast dish. Unfortunately, in recent years it has been degraded by "instantizing" and "flavorizing" until it no longer resembles oatmeal.

There is, however, a wonderful tin of Irish oatmeal that may be purchased if one hunts diligently enough in the specialty grocery stores. It is made in Kildare, Ireland from irish oats and labelled McCann's Finest Oatmeal. Made according to the recipe printed on the can it produces a steaming hearty breakfast or supper dish with a pleasing, nutty texture and flavor.

"Into 4 cups briskly boiling water sprinkle one cup oatmeal and 1 tsp. salt, stirring well. When porridge is smooth and beginning to thicken, reduce heat and simmer 30 minutes, stirring occasionally. May also be made in double boiler. Avoid overcooking to insure nuttiness."

Serve with cream and sugar, or butter and hot milk, as the Scots do.

A less expensive, but quite good substitute is "steel-cut" oats, now carried in most health food stores.

FRIED PUDDING

7½ cups water   1½ tsp. salt
2 cups cornmeal

Many years have passed since cornmeal mush bubbled in an iron kettle suspended over a hearth fire, but it may still be made in an open kettle, and it tastes best that way.

Stir the cornmeal into a very heavy steel or aluminum kettle to prevent it from burning on the bottom. Bring mixture to a boil over moderate heat, then turn heat back to low and simmer slowly for about 1 hour. Stir often using a long-handled spoon so the bubbling hot meal will not spatter onto your hand. Mush should be fairly thick and taste deliciously sweet, not raw, when done. (If you're impatient, you can speed things up by using a pressure cooker. Gradually stir cornmeal into boiling salted water in pressure cooker. Place cover on cooker. Allow steam to flow from vent

pipe to allow air to escape from cooker. Place indicator weight on vent pipe, adjust heat and cook 10 minutes. Allow cooker to cool.)

Pour corn meal mush into buttered 9″ square pottery or glass dish. Cool, uncovered. Refrigerate overnight.

Slice in 1/4″–1/2″ slices and pan fry slices in butter over low to medium heat until brown and crisp on both sides (takes a while). Serve with molasses or maple syrup.

You might even enjoy a serving of the warm mush with milk and a light sprinkling of sugar for a warm supper dish as was the custom in former days.

With two or three cups of steaming Mocha and Java coffee—a favorite New England blend—and perhaps a few slices of crisp bacon, this makes a breakfast that will "stick to your ribs" on a zero day.

Modern tastes may prefer maple syrup, but this dish has always been served at Arrowhead with molasses, and we still prefer it that way. Molasses was a traditional sweetener in New England, and was especially available in Newburyport because of the active West Indies trade here. Charlotte remembers, as late as the early 1920's, going to "Tumps" Bowlen's store on Merrimac Street with her father and seeing molasses pumped from the barrel into the family molasses jug.

This is one of the easiest—and tastiest—recipes for breakfast bread we know.

BROWN BREAD GEMS
(ALSO KNOWN AS BACK AND FORTH CAKES)

½ cup rye flour   ½ tsp. salt
½ cup white flour   1 tsp. soda
1 cup Indian meal   ½ cup molasses
                1 cup milk

Dissolve soda in molasses. Sift other dry ingredients together; stir in molasses and milk. Pour into greased gem pan* or muffin pan. Bake at 400° twenty minutes.

Iron gem pan

MOLASSES MUFFINS

2 cups flour   ½ tsp. ginger
3 tsp. powder   ¼ cup shortening
3 tbsp. sugar   1 egg, unbeaten
½ tsp. salt   ⅓ cup molasses
                ¾ cup milk

Sift dry ingredients together. Cut in shortening with a knife until size of peas. Make a well in center and add egg, molasses, and milk all together. Stir until just mixed. Fill well-greased muffin pans 2/3 full. Bake at 400° 15–20 minutes.

These muffins, which use the traditional combination of molasses and ginger, are newcomers to our cookbook. Being made totally with white flour, they are light and smooth textured.

*A gem pan was standard equipment in a New England kitchen. Made of heavy iron, it had twelve half-cylinder sections. When the batter rose, it formed a flattened cylinder shape reminiscent of the old-fashioned, pre-brilliant-cut shape of gems. The "gems" also rolled back and forth on their round bottoms; hence, their nickname.

CHRISTMAS AT ARROWHEAD

When folks at Arrowhead started to observe Christmas we're not sure, but by Charlie's youth in the 1880's it seems to have been well established. An orange and a piece of silver money in the toe of the stocking were a tradition in his childhood. (The same tradition has also come down in Paula's family.) We also have a number of Christmas issues of his *Youth's Companion* magazine, to which he subscribed, showing typical celebrations in that era; and a charming picture of Saint Nicholas in his sooty nineteenth-century guise which is still brought out some years to preside over the festivities.

The formal celebration of Christmas in the early twentieth century began with the Belleville Church Christmas Fair, two or three weeks before Christmas. Here you could buy presents which the Home Circle ladies had made: dolls and doll clothes, pretty aprons, pot holders, pen wipers, mittens and scarves, crocheted doilies and antimascassars. At home people made presents for each other, too, with great secrecy and conspiracy.

A few days before Christmas came the great Christmas tree expedition. All year, Charlie and others had sized up and noted the location of likely looking cedar trees. Often, however, when you got to them the one near the pasture gate was too thin, and the one near Will's fence was too brown, and the one down the hill from the old hen pen had a double top. Eventually, after much tramping back and forth in the snow, we found one that would do, cut it, and dragged it back to the house, where hot cocoa would revive us. Another cold trip, this time to the attic, produced the familiar cardboard box with its tiny red-and-green diamond pattern, containing a venerable assortment of ornaments. Greens, pine cones, and berries from the place provided small wreaths for the windows, and decorations for mantels, mirrors, and picture frames.

The Day itself was a happy round of stockings, presents, and delicious food on the best china, silver, and cut glass, served on the heavy white damask table cloth with its matching two-foot-square, crisply ironed napkins. After dinner, and during the week before New Year's Day, everyone's presents would be displayed in the parlor, and visits would be exchanged with friends to admire them.

While Justin today gets his full share of electronic games and battery-powered whiz-bangs, Christmas at Arrowhead still retains much of the old flavor: a tree cut on the place, a princess pine wreath on the door, Christmas pudding with the hard sauce in the cut glass hard sauce dish—and even the same jokes about who is being piggy with the hard sauce.

John Currier Moulton playing Father Christmas, 1904. LMP

CHRISTMAS BREAKFASTS

#1
Grapefruit halves
Chicken pies
Toast
Coffee

#2
Grapefruit halves
Boneless sirloin steaks
(Seasoned with salt, pepper, and
a pinch of crushed garlic)
Scrambled or poached eggs
Blueberry muffins

CHRISTMAS MENU #1

Roast turkey with cracker stuffing
Mashed potatotes with gravy
Prince Edward Island turnips
Squash
Creamed onions
Cranberry sauce
Celery
Parker House rolls
Fruit bread*
Christmas Pudding with hard and soft sauces
Cranberry, mince or squash pie
Penuche, fudge
Coffee

CHRISTMAS MENU #2

Rolled sirloin roast
Roast goose with fruit stuffing
Mashed potatoes
Turnip
Peas
Scalloped oysters
Candied yams
Cranberry sauce
Baked creamed onions
Fruit bread
Christmas pudding
Pies—squash, cranberry
Tapioca pudding

CHRISTMAS BREAKFAST CHICKEN PIE

Prepare chicken as for Chicken Fricasee. Remove bones and place small pieces of meat in buttered ironstone baking dish. Slightly thicken stock from kettle with a little flour and cover chicken with it. Refrigerate overnight.

Before breakfast make a shortcake as for strawberries. Roll out to about 1/2″ thickness and completely cover chicken in baking dish, making slash in middle. Bake in 400° oven about 1/2 hour.

Shortcake crust is the real old-fashioned approach to chicken pie and your family will need no coaxing to Christmas breakfast, or any other meal at which this is served.

ROLLED SIRLOIN ROAST

Recent years Christmas Dinner at Arrowhead has been made easier with the addition of a Jenn-Aire Range at Dick and Paula's house. While the goose is roasting to perfection in the convection oven, the beef is merrily cooking on the range-top rotisserie. A 3–3 1/2 pound roast will take about 2 1/2 hours this way. A little fresh ground pepper is the only seasoning needed for a perfect, juicy roast.

Polly stirring the Christmas pudding, 1980's. CMC

ROAST GOOSE

Allow app. 1 1/4 lbs. per person. Wash and pat dry, rub with salt and pepper, stuff loosely and fasten with skewers. Prick skin well. Place on rack in roasting pan. Put in preheated 425° oven for 20 minutes, pour off fat and reduce oven to 300° to finish cooking, app. 20 min. per pound. Baste often with cider.

STUFFING — PREPARE DAY BEFORE

1 lb. sausage meat
2 cups dry bread crumbs
1 lg. onion, chopped
2 stalks celery, chopped
½ cup dried apricots
½ cup dried pears
½ cup prunes
½ cup dried peaches

1 cup nuts (pecans, walnuts)
3 apples — cored, peeled and
  diced
salt and pepper
½ tsp. dried crushed sage
bay leaf
1 tbs. chopped parsley
cider or wine

Soak dried fruit in cider or wine, chop fine. Cook sausage until browned, remove to large bowl. Cook onion and celery lightly in sausage fat. Add to sausage meat with breadcrumbs, chopped fruit, nuts, and seasoning. Cover and refrigerate overnight. Check for seasoning and add cider if too dry in the morning.

HOLIDAY SCALLOPED OYSTERS

2 cups crushed common crackers
1 qt. shucked oysters with
  liquid

½ tsp. Worcestershire sauce
1 tbsp. sherry
¼ lb. butter

2 cups half-and-half milk/lt. cream

Layer crackers, oysters, dots of butter, very little liquid; add milk, Worcestershire sauce and sherry. Let set overnight. Bake 350° for 1 hour. (Before baking, add more milk to top of layers if needed.)

BAKED CREAMED ONIONS

Boil 1 lb. small white onions until soft. Put into 2 qt. casserole. Make a white sauce of 4 tbsp. flour, 4 tbsp. butter, 2 cups whole milk. Add to sauce 1/2 cup grated sharp white cheddar. Pour sauce over onions in casserole. Cover with buttered bread crumbs made of 2 slices whole wheat and 2 tbsp. butter. Cook in 350° oven until bubbly hot.

TAPIOCA PUDDING

6 tbl. minute-type tapioca	4 cups milk
6 tbl. sugar	1½ tsp. vanilla
3 eggs—separated	4 tbsp. sugar

Mix together in saucepan tapioca, sugar, egg yolks, and milk; let stand 5 minutes. Beat egg whites until foamy, gradually beat in 4 tbsp. sugar. Set aside. Cook tapioca mixture over medium heat to full boil, stirring constantly. Remove from heat and gently fold into egg white, stir in vanilla. Serve warm or chilled.

AUNT LIZZIE'S MOLASSES FRUIT CUPCAKES

1 cup butter	4 cups flour
1 cup sugar	1 tsp. cinnamon
2 eggs, beaten	1 tsp. nutmeg
1 cup molasses	½ tsp. allspice
1 teaspoon baking soda	½ tsp. cloves
1 cup milk	1 lb. citron

1 lb. currants

Cream butter and sugar; add eggs. Dissolve soda in molasses and beat into sugar mixture. Add milk alternately with flour which has been sifted with spices. Fold in lightly floured currants and citron. Spoon into greased cupcake tins, and bake at 375°, 20–25 minutes, or until firm to touch. Store in tightly covered tins and allow to "ripen" for several days for best flavor.

HEARTY WINTER MAIN DISHES

This is an old recipe, dating from the days when such concoctions were more often called "meat pudding" than "meat loaf."

MEAT PUDDING

1½ lbs. ground chuck
2 common crackers, crumbled
2 slices bread, crumbled
1 egg, beaten
¾ cup milk
1 medium sweet pepper,
  chopped fine
1 medium onion, chopped fine

2 tbsp. catsup
½ tsp. Lea and Perrin's
  Worcestershire sauce
1 tsp. Bell's seasoning
1 or 2 leaves fresh or dried sage
½ tsp. salt and sprinkling of
  ground pepper

Mix crumbled crackers and bread into ground meat. Add beaten egg. Mix in chopped pepper and onion. Add milk and seasonings.

Spread several tbsp. tomato or sweet pepper relish or catsup in bottom of buttered baking dish and put meat mixture on top. Dot with butter. Place in preheated 400° oven, turn temp. back to 375° and bake about 1 hour. Top should be browned and interior soft.

CHARLOTTE'S MEAT LOAF

2 cups soft bread crumbs
2 tbsp. melted butter
1½ lbs. ground beef
½ cup catsup

½ cup water
1 egg well beaten
2 tbsp. onion, finely chopped
1 tsp. salt

1 pkg. sm. frozen peas, or 1 can sm. peas

Toss 1 1/2 cups bread crumbs with melted butter, and set aside. Combine rest of crumbs with the remaining ingredients except the peas, mixing lightly with fork. Pack mixture loosely in glass oven-bake dish 8" x 1 1/2". Lightly press buttered crumbs on top. Bake in 350° oven for 1 hour. Cut into servings and arrange around edge of warm serving platter. Put cooked peas in center and garnish with tomato or fresh orange slices. Serves six.

For those of you who still enjoy using your pressure cooker, we print this delicious recipe.

PRESSURE COOKER MEAT LOAF

2 lbs. ground beef	*1 medium carrot, grated*
1 tsp. salt	*½ cup corn flakes*
fresh ground pepper to taste	*2 tbsp. catsup*
2 eggs slightly beaten	*½ tsp. Worcestershire Sauce,*
1 medium onion, finely diced	*1 tbsp. shortening*
1 stalk celery, finely chopped	*2 tbsp. water*
4–5 medium small potatoes, peeled	

Have beef ground twice. Season with salt and pepper. Add eggs, onion, celery, carrot, corn flakes, catsup, and Worcestershire. Mix thoroughly, form into two small loaves, wrap in wax paper and refrigerate for 2–3 hours.

Heat cooker and add shortening. Brown each loaf on all sides, turning with spatula. Place loaves on trivet and potatoes around them. Follow directions for your pressure cooker, allowing 15 minutes cooking time, and allowing cooker to cool before removing meat to platter. To juice remaining in cooker, add 1 cup water and 2 tbsp. flour. Cook, stirring constantly, until thickened. This will make a pale but delicious gravy.

PAULA'S POT ROAST

3–4 lb. pot roast	*2 small turnips*
6 carrots	*2 stalks celery*
2 medium yellow onions	*1 clove garlic, mashed*
2–3 red tomatoes	*dry parsley*
1 green sweet pepper	*dry red wine*

Brown roast (bone in) in butter. (A 3–4 lb. roast feeds four easily.) Dice finely carrots, onions, tomatoes, sweet pepper, turnips (white *turnips*; not rutabaga), celery. Add to roast with garlic, parsley—generous, and dry red wine—very generous. Cook covered, very slowly 2–3 hours. Salt and pepper to taste at the table.

STANDING RIB ROAST

Buy the first 2 ribs of a really prime roast. Pat flour into fat on top of meat and stand on ribs in open shallow roasting pan. Use no water. Place in preheated oven 450° for 15 minutes to seal in juices. Then lower heat to 325° and check roasting chart weight time and temperature for meat as you desire it.

For a 6 lb. rib roast Arrowhead roasts at 450° for 15 minutes. Then about 1 3/4 hrs. longer at 325°. This produces roast nicely brown outside with medium rare inside slices.

For gravy, work 4 tbsp. flour into 1/2 cup cold water to form smooth paste. Stir in 3 cups more water and add to browned fat. Simmer 5–6 minutes in roast pan, stirring constantly.

ROAST BEEF PIE

Cut scraps and chunks of beef from your leftover rib roast. Cook till barely soft a few chunks of carrots and potatoes, a medium onion coarsely chopped, and 1/2 pkg. frozen peas. Thin down leftover beef gravy with water enough to cover and pour it over mixture. Top with baking powder biscuit crust, as for shortcake, but rolled to 1/2-inch thickness, and slashed. Bake in hot oven 400° about 30 minutes. When done, crust should be brown, and inside bubbly.

POTATOES

Along with most farms in the Northeast, Arrowhead considered potatoes a staple for many years. Well adapted to New England's relatively poor soil, dependably productive, easy to store, and versatile in cooking, they were a widely popular crop. Charlie was well known for his crops of Green Mountains in the early years of this century.

In recent years, however, changing economies and cultural practices have altered all this. Potatoes are now a crop more suited to large scale production. They require a large investment in specialized machinery for planting, harvesting, and marketing. They also require use of more specialized insecticides, as potato beetles have developed resistance to earlier pesticides.

Because of this, we no longer grow this crop at Arrowhead. Instead, we buy potatoes from larger growers in the area. Our favorite winter keepers are Green Mountains and Russets. Earlier in the year we offer the pink-skinned varieties—Norland, Bliss, and Early Rose. All of these are of extra nice eating quality. You can, of course, still grow your own. It requires some hard labor and also some diligent hand picking of beetles. But if you're a gardener, your own new potatoes are well worth the effort.

Harking back to the days when potatoes, cabbages, winter squashes, and root crops were the only available winter vegetables, many winter dinners and suppers still "taste right" with these, even now that almost any vegetable is available all winter.

SCALLOPED POTATOES

Line the bottom of a well-buttered baking dish with thinly sliced potatoes, Green Mountains if available, making layer 3/4 inch deep. Sprinkle with bit of salt and fresh pepper, add a few pieces of sliced onion and dots of butter equalling a tablespoon. Run flour sifter over it lightly. Continue thus for 3 or 4 layers. Then pour in enough milk to just cover top layer of potatoes. Bake in 375° oven for about 45–60 minutes depending on depth of potato layers. Top should be nicely browned and bubbly when done.

If you are a purist, try this without the onion. For an all-in-one supper dish, layer some leftover baked ham or pork shoulder in it.

HASHED BROWN POTATOES

Boil eight or nine medium potatoes, preferably Green Mountain. When cold, chop into 1/2" bits; add salt and pepper to taste. Melt two tablespoons butter in fry pan. When hot, put in potatoes and stir with fork to coat with butter, then spread out evenly, turning heat back to medium. Let brown on bottom before turning with spatula. Lower heat a bit more and let other side brown very slowly. Slow browning gives them a crispy little coat. Hold on warm till wanted.

A little cold roast beef or cooked lamb ground fine may be added before potatoes finish browning, also a little gravy, if you wish. Either way, hashed browns are good to the last bite.

FRIED POTATO CAKES

Mashed potatoes are the base for these delicious breakfast, dinner or supper cakes. So whenever you run the potato masher, make enough to have some left over. Mould the mashed potato into little round cakes about 1" thick and dip each into a bowl of beaten egg; quickly roll in a platter of flour and drop into a hot buttered fry pan. Using medium heat, brown the cakes, adding more butter as necessary. This takes a while, so be patient.

These are good in the morning with bacon and eggs, as a hot touch with cold luncheon meat or for a light supper with sausage and cold applesauce.

PARTY POTATOES

8–10 cooked, hot potatoes	*1 tsp. garlic salt*
1 cup sour cream	*1 tsp. salt*
8 oz. cream cheese	*butter and paprika*

Beat sour cream and cream cheese until blended. Add hot potatoes, one at a time, beating constantly. Spoon into 2 qt. casserole. Refrigerate 24 hours. Dot with butter, sprinkle with paprika. Bake at 350° for 45 minutes.

POTATO SOUP

5 medium potatoes, boiled	*bit of salt and fresh pepper*
2 tbsp. finely chopped onion	*3 tbsp. butter*
2 tbsp. finely chopped carrot	*2 tbsp. flour*
¼ tsp. celery salt	*4 cups milk*

Cook onion and carrot in butter 5 minutes; add flour which has been smoothed into a paste with a little cold water, add milk and seasonings. Cook in double boiler 20 minutes. Put hot boiled potatoes through masher and stir all together. Strain and serve in pottery soup bowls garnished with a few sprigs of chopped chives from your window sill.

STEAMED CABBAGE

With very little work common cabbage becomes a very classy vegetable. Cut through the center of cabbage with sharp heavy knife and continue cutting each half into 1/2-inch slices. Rinse with cold water and place in large covered fry pan with 3 tbsp. butter. Put burner on high long enough to heat pan, then switch to low heat and steam cabbage 5–8 minutes, depending on texture.

Salt lightly and serve this crispy wonder immediately with butter and pepper, or vinegar.

SOUPS, STEWS AND CHOWDERS

Lamb, which has always been so much a part of New England cooking, is the base for this nourishing soup, and gives it a distinctive flavor. It can be made with beef, but it will taste different. The vegetables and their proportions are only suggestions; use whatever you have.

POLLY'S VEGETABLE SOUP

*1–2 lb. stewing lamb or lamb
  bones with some meat attached*
bay leaf
salt
freshly ground pepper
1 large can whole tomatoes
2–4 carrots

2–3 stalks celery
3–4 small onions
1 cup lima beans, dried or frozen
3–4 small potatoes
1 cup green beans
½ cup green peas
½ cup kernel corn

Place lamb in large soup kettle with two or three bay leaves and enough water to cover completely. Season with salt and freshly ground pepper. Bring to a boil and simmer slowly until meat falls away from bones —1 or 2 hours. Cool. Place in refrigerator for several hours to congeal fat. Discard bay leaf, fat, and bones; cut meat finely and reserve. To broth, add two new bay leaves and canned tomatoes with their liquid, and bring to simmer. If you are using dried lima beans, add them now and simmer for about an hour, or until they are partially soft. Then add two or three large stalks of celery, sliced, and two to four carrots (depending on size), peeled and sliced. Next, while these are simmering, peel and slice three or four smallish onions and add these. While these are simmering, peel and slice three or four small potatoes and add. When these vegetables are nearly tender, add reserved meat and frozen lima beans if you are using them. When broth has returned to simmer, add green beans. When it returns to simmer again, add peas and corn, and adjust seasoning. As soon as soup returns to simmer, turn off heat and allow to sit for a few hours to gather flavor. Serve piping hot with Indian Cake or another hearty homemade bread and butter. This will keep—even improve—for several days in the refrigerator, but be careful not to overcook the vegetables when reheating this soup.

PAULA'S BEEF STEW

2 lbs. stew beef	1 tbsp. salt
1 tsp. Worcestershire	1 tsp. sugar
½–¾ cup red wine	1 tsp. paprika
1 clove garlic	¼ tsp. pepper
1 medium onion, sliced	dash alspice
2 bay leaves	vegetables

Brown meat in Dutch oven in shortening. Add hot water (about 4 cups), red wine (burgundy or chianti), and next 9 ingredients. Cook, covered, very slowly for 1 hour. Add potatoes, carrots, yellow turnips, small white onions, and cook until vegetables are done. Remove bay leaves. Best if made a day ahead.

CHOWDERS

Fish chowder is one of the oldest New England dishes. It was probably introduced as *chaudier* by the French and other European fishermen who camped out on New England's shores during the summer in the sixteenth century to cure the fish they caught on the Grand Banks. It can be made with many kinds of fish including cod and flounder, but we have always preferred haddock. While we have heard that certain areas in New England actually did favor thick chowders, Newburyport was not one of them, and these recipes reflect our preference for a milky chowder with firm, distinguishable "solids."

FISH CHOWDER

Place 2 1/2 lbs. fresh haddock fillets in kettle and simmer gently in 1 quart unsalted water for 5–10 minutes, or until skin can be peeled off. Remove fish and add 1 medium onion, coarsely chopped. Simmer 5 minutes. Add 4–5 medium potatoes, pared and diced coarsely. Continue to simmer until both are barely soft. Add 1 quart milk, already scalded.

Put fish back into liquid in good-sized pieces. Simmer together gently 5 or 6 minutes, season to taste with salt and fresh ground pepper, turn heat back to warm and let set 2–3 hrs. to gather flavor. *Do not let it boil.* Just before serving, add 2–3 tbsp. butter and 10 common crackers which have been split and soaked in enough cold milk to cover. Let chowder set 10 more minutes over increased heat. Serve piping hot in chowder plates with a tsp. butter melting in each. Prepare for second helpings.

Paula, for a low-sodium diet, substitutes tarragon for salt and the result is delicious.

The Indians dug clams in the mud flats near the mouth of the Merrimack long before the white man came, and by the eighteenth century fishing and clamming were such well-established industries in the area "downalong" the river that it was called Joppa, for the Biblical fishing village. (Newburyporters pronounce it "Joppy.") After being nearly ruined by pollution, the clam flats are now open again and producing the tender, sweet little clams for which the area has long been famous.

CLAM CHOWDER

4 cups clams, fresh from the
   flats, shucked
4 cups sliced potatoes

2 medium onions, chopped
5 cups milk, scalded
salt and fresh pepper to taste

4 tbsp. butter

Remove black skin and bags from clams and set aside, saving all liquor.

Chop onions and simmer in a little water for 3 minutes. Add sliced potatoes and just cover with water, simmering over low heat until barely soft. (If cooked over high heat, chowder is apt to stick to bottom of kettle.) Add clams, and after they come to a boil, simmer 2 minutes only. Longer cooking will toughen them. Remove from heat and let set for 5–6 minutes, then add hot scalded milk, clam liquor and seasoning. Stir in butter. Let chowder set on warm burner 2 or 3 hours to gather flavor, never letting it boil.

Serve hot with crisp little crackers. This is a sweet buttery chowder and will be even better tomorrow—if it lasts that long.

OYSTER STEW

1 pint large oysters
5 cups milk

1 tsp. salt
freshly ground pepper to taste

2 tbsp. butter

Scald milk in heavy kettle. Place oysters and juice in fry pan over low heat. Check for bits of shell. Cook until edge of oyster is barely fluted. Then combine with milk, salt, dash of pepper, and butter. Let set an hour with heat adjusted so stew is just under the simmering point. Stir occasionally. Serve piping hot with generous dabs of butter and oven-freshened oyster crackers.

WINTER SUPPERS

Baked beans on Saturday night is one of New England's oldest culinary traditions. The Puritan prohibition against doing any work on Sunday included cooking; therefore, the Sunday meals had to be prepared on Saturday. The baked beans for Saturday supper required little effort other than an occasional stir and addition of water, while Sunday's meals progressed.

High in protein and easy to store all winter, beans formed an important part of the earliest settlers' diet, as they had for the Indians, and they have continued to be popular ever since. Charlotte's father, Charlie, not only ate baked beans and brown bread every Saturday night of his eighty-seven years, but he was always greatly disappointed if there were not enough left over to be reheated for Sunday morning breakfast! While succeeding generations have never been that enthusiastic about them, we have always been very attached to having them on Saturday nights, and have even developed the Saturday night bean picnic as an institution, with the bean pot nestled in a "cozy" inside the picnic basket.

BAKED BEANS

2 *cups beans (yellow eye,*	¼ *cup molasses*
kidney, or pea)	2–3 *slices bacon*
2 *tsp. dry mustard*	1 *tsp. salt*

Soak beans in cold water till they wrinkle (overnight). Place in beanpot with all other ingredients and water to barely cover. Bake at 300° 4–5 hours, or 5–6 hours at 275°, depending on age and consistency of beans. Stir frequently, adding water to barely cover at each stirring—do not allow to dry out. Serve with Red and Green Tomato Relish, or Aunt Ida's Shirley Sauce, and hot buttered brown bread.

BAKED BROWN BREAD

1 egg
⅓ cup sugar
½ tsp. soda
⅓ cup molasses
1¾ cup All-Bran cereal

⅔ cup Indian meal
¼ cup melted shortening
2 cups flour
2 tsp. baking powder
½ tsp. salt

1½ cups milk

Beat egg and sugar together; add soda dissolved in molasses, All-Bran, Indian meal, and shortening. Sift flour, baking powder, and salt, add to mixture alternately with milk. Bake in greased tin at 350° for 1 hour, 15 minutes.

This modernized recipe is the one we have used at Arrowhead for the past forty years. Sarah introduced it when graham flour for the earlier favorite recipe became unavailable, and now that it's available again in natural food stores, we continue with this version. We still bake it, of course, in the traditional round brown bread tin, but it tastes equally good baked in a loaf pan. It makes a firm, moist bread which is delicious cold as well as hot, and makes wonderful toast.

FINNAN HADDIE BROILED OR BAKED

Finnan Haddie — salted, smoked haddock, named after the Scottish town of Findon, which was famous for it — is an old, old New England favorite on the winter supper table. On a cold January night, the lids would be removed from the iron range and a wire hand broiler holding a finnan haddie steak would be placed over the glowing coals, broiled 10 minutes on the first side, turned and broiled 8 minutes longer. It was then placed on an ironstone platter and dotted with butter. A rich aroma rose from the smoked fish and the hanging brass lamp shed a soft glow on the family gathered around the big kitchen table. Baked potatoes were served with the fish, and fruit from the preserve cellar.

Sometimes the finnan haddie was baked. For this, soak fish in cold water 10 minutes, drain and rinse, place in shallow baking dish and barely cover with milk. Bake in 375° oven 30 minutes or until sufficiently browned. Take to table with butter melting over it and enjoy its mild sweet flavor.

Smelts have always been a popular dish in this part of the country. Smelts used to be netted during spawning runs in the spring, as well as caught through the ice in winter but this is no longer allowed as it was endangering the smelts. They are still caught by line in the winter, however. As soon as the ice is thick enough smelt houses begin to appear on it, providing cozy shelters for the fishermen.

FRIED SMELTS

When your special fisherman brings home a fine catch from Great Bay or Parker River he hopes for a tasty meal of fried smelts. If he is really a good sport he will clean them while you beat the egg to dip them into and get out the corn meal to roll them in. Now drop them into a "hizzling hot" frypan, turn down the burner and cook very slowly about 15 minutes to first side and 10 minutes to the other. This makes them crispy outside and moist within.

Happy eating!

INDIAN CAKE

This is an adaptation of one of Aunt Ida's recipes, and is one of our all-time favorites. As quick to fix as most mixes, it makes a hefty, delicious cornbread with an authoritative crust. Moist but crumbly when hot, it "sets up" nicely when cold. Serve it hot in large squares, with bacon or sausage and rosy homemade applesauce, for a wonderful Sunday supper. Slice it thin and serve cold as a hearty picnic bread to accompany cold sliced meat. Or slice thin and toast it until its edges brown for some of the best toast you every spread butter and jelly on. Stone ground corn meal is particularly good for this recipe.

2 cups Indian meal	1 tsp. baking soda
2 cups white flour	2 tsp. cream of tartar
½ cup sugar	2 cups milk
½ tsp. salt	1 egg
cooking oil to fill egg shell	

Place Indian meal in large mixing bowl; sift in flour, sugar, salt, soda, and cream of tartar. Break egg into mixture. Fill half of egg shell twice with salad oil and add. Add milk. Mix with spoon or electric mixer until thoroughly blended.

Pour into greased 9" baking pan. Bake at 400° 45 minutes, or until straw inserted in center comes out clean. You may have to lower heat a little at end of baking time to prevent over-browning.

ELIZABETH'S SUNDAY SUPPER CASSEROLE

10 slices bread
½ lb. sharp cheese
3 eggs

2 cups milk
dash dry mustard (about
½ tsp.)

salt and pepper

Trim crusts from bread; butter lightly, then cut in cubes. Grate cheese.

Beat together eggs, milk, dry mustard, and salt and pepper to taste. Line greased casserole with bread cubes. Then sprinkle on some of the cheese; then more bread cubes, followed by more grated cheese. Finish off with bread. Pour liquid mixture over all.

Use sharp Cracker Barrel or some not-too-milky cheese. Let stand 3–4 hrs. before baking. Bake 275° for about 45 min. to 1 hr. Top should be lightly browned and custardy part should be done then. You can also add diced or cubed ham bits or tuna fish to make a heartier dish. Also, grilled franks go well with it.

MOLASSES AND SPICE FAVORITES

The combination of molasses and ginger was one of the most popular flavors in New England cooking, and gingerbread existed in many variations. This is one of our favorites. Like many recipes using molasses, it calls for "saleratus" or baking soda as the rising agent.

SPONGE GINGERBREAD

½ cup sugar
good ⅓ cup butter
1 egg
1 cup molasses

2 cups flour
1 tsp. ginger
¼ tsp. salt
1 tsp. baking soda

1 cup boiling water

Cream butter and sugar; beat in egg and molasses. Sift flour with ginger and salt; stir gradually into molasses mixture. Dissolve soda in boiling water and add last. Pour into well-greased 7" x 11," or 9" square pan. Bake at 375° for 25 minutes.

Polly's Double Gingerbread: Add 1–2 tsp. finely chopped preserved ginger with flour.

Big, soft molasses cookies have always been a staple of New England childhood. Our old cookbooks are full of recipes for them, but this is our favorite. It came to Arrowhead from Glen's mother, "Pink" (Eliza Woods Chase, 1888–1967), whose maternal grandfather, Robert Bayley, was the last West Indies merchant to import molasses into Newburyport. Bayley's Wharf, near the foot of Fair Street, was piled high with barrels of molasses until it closed in the 1870's. The memory lingers on in these tasty cookies, which are wonderfully easy to make, and store well.

NANA PINK'S MOLASSES COOKIES

½ cup molasses	1 tsp. salt
½ cup sugar	2½ cups flour
1 egg, beaten	1 tsp. soda, dissolved in
⅓ cup melted shortening	½ cup boiling water
½ tsp. ginger	

Combine boiling water, molasses, and sugar in large bowl. Add egg and shortening. Sift dry ingredients and add to molasses mixture. Drop by large spoonfuls on greased cookie sheet. Bake at 375° 15–20 minutes.

GINGER SNAPS

¾ cup shortening	2 tsp. soda
1 cup light brown sugar	½ tsp. salt
¼ cup molasses	1 tsp. ground ginger
1 egg	1 tsp. ground cinnamon
2¼ cups flour	½ tsp. ground cloves

Cream shortening, brown sugar, molasses, and egg until fluffy. Sift together flour, soda, salt, ginger, cinnamon, cloves, and stir into molasses mixture.

Form into little balls, roll in granulated sugar. Press down lightly on greased cookie sheet, bake in 375° oven for 12 minutes.

Hermits are among the most traditional of New England dessert treats. They combine with native farm products many of the ingredients which came in with the West Indies and Mediterranean trades. They also pack, store, and travel well, and we suspect that Uncle Rufus Wigglesworth, as well as many other New England sailors, was often supplied with a cache of these goodies from home in his sea chest as he began a long voyage.

HERMITS

6 tbsp. shortening   1½ cups flour
1 cup lt. brown sugar   ½ tsp. cloves
1 egg   ½ tsp. allspice
½ cup milk   1 tsp. cinnamon
2 tsp. baking powder   ½ cup seedless raisins
             ½ cup chopped nut meats

Cream shortening, add sugar and beaten egg. Mix well together. Sift dry ingredients together and add alternately with milk. Add lightly floured raisins and nuts. Drop from spoon onto buttered cookie sheets and bake in 375° oven about 15 minutes.

AARON'S BUNDLES

Why are these called Aaron's Bundles? They are bundle-shaped, but why "Aaron's"? We have been unable to determine, even after a diligent search of the Bible. Of course, many early New Englanders were named after Old Testament figures, so perhaps we are looking in the wrong place. Perhaps Aaron was a sailor or sea captain who brought home the spices used in them. Whatever the origin of the name, folks at Arrowhead obviously enjoyed these spicy little bundles—the recipe is the most spattered and tattered page in the 1901 *Belleville Church Home Circle Recipe Book*.

1 cup sugar   1 tsp. soda
½ cup butter   1¾ cups flour
3 scant tbsp. molasses   ½ tsp. cinnamon
1 egg   ½ tsp. ginger
1 cup milk, sweet   ½ tsp. nutmeg
             ½ cup raisins

Cream sugar and butter together; add molasses, and beaten egg. Dissolve soda in milk and add alternately with flour which has been sifted together with spices. Stir in lightly floured raisins. Bake in buttered gem pans in preheated 375° oven 20–25 minutes, or until firm to touch.

HEARTY WINTER PUDDINGS

Indian Pudding is one of the most traditional of New England foods, dating back to the first settlers. Deprived of their familiar English grains, they used the "Indian meal" as a substitute in many different mushes, porridges, puddings, breads and cakes. These were sometimes sweetened with whatever was at hand—honey, maple sugar, or molasses from the West Indies. As trade expanded, spices became available, and ginger and cinnamon became a standard part of "Indian Pudding."

INDIAN PUDDING

½ cup cornmeal
1 qt. milk, scalded
½ tsp. salt

½ cup molasses
2 tbsp. butter
½ tsp. ginger

½ tsp. cinnamon

In buttered ironstone or glass baking dish, pour scalded milk over cornmeal. Add rest of ingredients and stir. Place in preheated 350° oven. After 15 minutes, reduce to 325°. Bake 2 1/2–3 hours, stirring often. Serve hot with plain or whipped cream, or, if you prefer, vanilla ice cream.

COTTAGE PUDDING

1 cup sugar
butter size of an egg
2 eggs
1½ cups flour

1 tsp. cream of tartar
½ tsp. baking soda
salt
½ cup milk

vanilla

Beat together sugar and butter. Add beaten yolks of eggs and stir in. Sift together flour, cream of tartar, baking soda, and bit of salt. Add sifted dry ingredients alternately with milk. Fold in stiffly beaten whites of eggs. Flavor with a little vanilla.

Cottage puddings were originally baked in their own little fluted round tins with a hole in the center. If your pantry lacks this, pour into small size angel cake tin and bake 350° for 30–40 minutes, or until cake stops "singing." Serve with chocolate sauce as made for ice cream.

RICE PUDDING

⅓ cup rice   ½ cup sugar
1 qt. milk   ¾ raisins
                ½ cup cream

Combine rice, milk, sugar, and raisins in buttered dish and bake at 375°
2–2 1/2 hours, stirring frequently. Add cream and bake 1/2 hour longer.
    May be served hot, with or without plain or whipped cream, or cold.
Another variation is to top it with meringue, sprinkle with coconut, and
continue to bake until topping is browned.

This pudding is a real treat for chocolate lovers—rich and dark and crusty
with a sweet and foamy sauce.

CHOCOLATE BREAD PUDDING WITH EGG SAUCE

4 slices bread, crumbed   ⅔ cup sugar
3 cups milk, scalded   2 sq. baking chocolate
2 eggs, beaten   2 tbsp. butter

Put bread crumbs into buttered ironstone baking dish. Shave chocolate
into bits and melt in saucepan with butter. Mix together beaten eggs, milk,
and sugar. Add melted chocolate to milk and egg mixture and stir all
together into baking dish with crumbs.
    Bake in preheated oven 375° one hour. Serve hot with egg sauce.

FOAMY EGG SAUCE

Beat one egg until light and fluffy. Gradually add 1 cup granulated sugar
while you continue beating.
    Spoon over your dishes of pudding and enjoy the compliments.

QUEEN OF PUDDINGS

2 cups bread crumbs   4 egg yolks
4 cups milk   bit of grated lemon rind
1 cup sugar   butter size of an egg, melted

Stir all together in the order given and pour into ironstone baking dish
and bake for 1 hour in 350° oven. Let cool slightly before frosting.

FROSTING

4 egg whites   1 cup sugar
                 juice of 1 lemon

Whip egg whites until stiff and gradually add sugar, beating all the while. Lastly beat in lemon juice, and quickly spread on top of pudding. Return to oven 350° and cook until meringue is slightly browned.
     Serve warm or cold with a spoonful of grape or strawberry jelly.

A snowy day with no school is an excellent time to fry doughnuts. "Many hands make light work" (sometimes they make more work, but it's fun anyway).

DOUGHNUTS

3 tbsp. shortening   ¼ tsp. cinnamon
1 cup sugar   ½ tsp. nutmeg
2 eggs   4½ tsp. baking powder
4 cups flour   1 cup milk
                 1 tsp. salt

Cream shortening, add sugar gradually and cream together. Add beaten eggs. Add flour sifted with remaining dry ingredients alternately with milk. Roll out dough on floured board to 3/8" thickness and ply the doughnut cutter. Drop into preheated 375° deep fat fryer, using either pure lard or vegetable shortening. Fry till golden brown underneath, turn and brown other side, never turning more than once. Don't forget to fry some holes.

APPLE FRITTERS

2 apples   ¼ tsp. salt
1 cup flour   1 egg
1½ tsp. baking powder   ⅔ cup milk

Pare and core apples and cut into small chunks. Sift together dry ingredients; add beaten egg and milk; stir until smooth. Stir apples into batter and drop by small spoonsful into electric deep fat fryer preheated to 375°. Batter should be thin enough to form into odd irregular shapes.

Fry till light brown underneath, turn with fork and brown other side. Turn only once. Drain on paper and keep warm in oven. Serve warm with maple syrup for dessert during winter vacation week. The children will love them—and you.

Another favorite version of this is made with bananas, which came into New England from the Caribbean as early as 1800, and have been popular ever since.

Waiting lines always formed for these, children to the left, grownups to the right.

SARAH'S PEPPERMINTS

2 cups sugar   *⅛ tsp. cream of tartar*
⅔ cup water   *1 tsp. peppermint extract*

Combine and heat sugar and water. When boiling rapidly, add cream of tartar.

When syrup forms thread when dropped from spoon, remove from heat and add peppermint extract. Beat fast with spoon until it begins to grain. Then drop quickly onto waxed paper to cool and harden.

SARAH'S HOMEMADE MARSHMALLOWS

1 tbsp. gelatine, dissolved in   *1½ cups sugar,* stirred into
½ cup water   *½ cup water*
1 tsp. vanilla

Bring sugar and water to a boil and pour over gelatine. Let mixture cool until it begins to stiffen. Add vanilla. Beat with egg beater until very stiff. Pour into 6″ square pan thickly dusted with powdered sugar. When firm, dust powdered sugar over top and cut in squares. Spread squares out on platter and let stand overnight. Then hide them until your party is ready.

Newest of a long line of children's favorites to concoct in the kitchen is this strange-sounding but quite delicious drink. It's more fun if you start with whole beans and grind them in an old-fashioned coffee mill, but any coarse-ground coffee will do.

JUSTIN'S JELLY SHAKE

2 cups cold milk
2 tbsp. coffee, coarsely ground
2 tsp. sugar

2 tsp. jelly, preferably
*   strawberry or grape*
straws

Pour cold milk into two glasses. Put 1 tbsp. ground coffee into each glass and stir thoroughly; then add 1 tsp. sugar to each, stirring well to dissolve. Lastly put 1 teaspoonful jelly in each glass (possibly you'd like a bit more) and stir rapidly. Now wait patiently a couple of minutes for the coarse ground coffee to rise and float on top, then push your straws to the bottom of the glasses and happy sipping with your Friend!

OTHER WINTER DESSERTS

These yummy desserts date from the days when most households had a well-stocked homemade preserve cellar. But a good facsimile can still be made with apricots, peaches or strawberries from a can. Arrowhead used strawberries from the cellar.

FRUIT ROLY-POLIES

Open jar, separating berries and juice. Pour juice into saucepan, add 1/2 cup sugar with 1 tsp. flour worked into it. Bring to a boil and set aside. You should have at least 1 cup juice. If necessary, add a little water.

Make a shortcake mixture and roll into rectangle 1/2 inch thick. Cover with fruit and roll up as for jellyroll. With sharp knife, slice into servings about 1″–1 1/4″ thick. Place flat in buttered cake pan and pour the juice around them. Bake in 400° oven about 25 minutes or until they are lightly browned and juice is thick and bubbly. Serve hot with whipped cream.

This tasty cake, an innovation at Arrowhead about 1950, has been a favorite ever since. It has, in fact, supplanted cherry pie for the celebration of Washington's birthday. And, baked in two heart-shaped layers (for which reduce the baking time slightly and make extra frosting), it makes a very special Valentine.

CHERRY NUT CAKE

16 maraschino cherries sliced	*½ cup butter*
½ cup walnuts chopped	*1¼ cups sugar*
2 cups sifted flour	*4 egg whites, beaten stiff*
3 tbsp. baking powder	*½ cup milk*
¼ tsp. salt	*¼ cup cherry juice*

Cream butter and sugar together. Sift dry ingredients together and add to creamed mixture alternately with milk. Fold in stiffly beaten egg whites. Shake cherries and nuts in a little flour and stir in. Bake in 9″ square pan about 45 minutes at 350°.

Make confectioners sugar frosting with the cherry juice and sprinkle with a few more nuts.

JELLY TARTS

To make these teatime jewels you will need:

1 recipe pastry, rolled about 1/8th inch thick	*2 jars bright flavored, colorful jelly*

From your tinware select a round cookie cutter and a doughnut cutter of same diameter. Fill a cookie sheet with matching pairs of cut pastry rounds, and put into a preheated 400° oven. Bake about 12 minutes until they are creamy in color but not beginning to brown. Just before serving put a spoonful of jelly in center of each whole round, lightly press the doughnut hole circlet on top and place a trayful of these beauties on your buffet.

*Originally Joseph Moulton's silver shop,
this ell is now the Arrowhead Farm office.* PCH

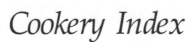

Cookery Index

ARROWHEAD FARM FROM MOULTON HILL, 1847,
IN THE BACKGROUND. A LITHOGRAPH BY BUFFORD